PLAIN LANGUAGE AND ETHICAL ACTION

Plain Language and Ethical Action examines and evaluates principles and practices of plain language that technical content producers can apply to meet their audiences' needs in an ethical way. Applying the BUROC framework (**B**ureaucratic, **U**nfamiliar, **R**ights **O**riented, and **C**ritical) to identify situations in which audiences will benefit from plain language, this work offers in-depth profiles to show how six organizations produce effective plain-language content. The profiles show plain-language projects done by organizations ranging from grassroots volunteers on a shoe-string budget to small nonprofits to consultants completing significant federal contacts. End-of-chapter questions and exercises provide tools for students and practitioners to reflect on and apply insights from the book. Reflecting global commitments to plain language, this volume includes a case study of a European group based in Sweden along with results from interviews with plain-language experts around the world, including Canada, England, South Africa, Portugal, Australia, and New Zealand.

This work is intended for use in courses in information design, technical and professional communication, health communication, and other areas producing plain-language communication. It is also a crucial resource for practitioners developing plain-language technical content and content strategists in a variety of fields, including health literacy, technical communication, and information design.

Russell Willerton teaches in the technical communication program at Boise State University. He has published articles on visual communication, white papers in high-tech industries, and online health information.

ATTW Book Series in Technical and Professional Communication

Jo Mackiewicz, Series Editor

Plain Language and Ethical Action: A Dialogic Approach to Technical Content in the Twenty-First Century
Russell Willerton

Rhetoric in the Flesh: Trained Vision, Technical Expertise, and the Gross Anatomy Lab
T. Kenny Fountain

Social Media in Disaster Response: How Experience Architects Can Build for Participation
Liza Potts

For additional information on this series, please visit http://www.attw.org/publications/book-series, and for information on other Routledge titles, visit www.routledge.com.

PLAIN LANGUAGE AND ETHICAL ACTION

A Dialogic Approach to Technical Content in the Twenty-First Century

Russell Willerton

Routledge
Taylor & Francis Group

NEW YORK AND LONDON

first published 2015
by Routledge
711 Third Avenue, New York, NY 10017

and by Routledge
2 Park Square, Milton Park, Abingdon, Oxon, OX14 4RN

Routledge is an imprint of the Taylor & Francis Group, an informa business

© 2015 Taylor & Francis

Library of Congress Cataloging-in-Publication Data
Plain language and ethical action : dialogic approach to technical content in
    the 21st century / Russell Willerton.
        pages cm. — (ATTW series in technical and professional communication)
    Includes bibliographical references and index.
    1. English language—Technical English.   2. English language—
Rhetoric.   3. Technical writing—Moral and ethical aspects.   4. Report
writing—Moral and ethical aspects.   I. Title.
    PE1475.W55 2015
    808.06'6—dc23
    2014043135

ISBN: 978-0-415-74105-7 (hbk)
ISBN: 978-0-415-74104-0 (pbk)
ISBN: 978-1-315-79695-6 (ebk)

Typeset in Minion
by Apex CoVantage, LLC

Printed and bound in the United States of America by Publishers Graphics,
LLC on sustainably sourced paper.

To the memory of my paternal grandmother, Josephine Willerton,
a remarkable, wonderful woman of letters and the arts.

CONTENTS

FIGURES AND TABLES

Figures

Tables

SERIES EDITOR FOREWORD

This book, Russell Willerton's *Plain English and Ethical Action*, is the third in the ATTW Book Series in Technical and Professional Communication (TPC). Like its predecessors, Liza Potts's *Social Media in Disaster Response: How Experience Architects Can Build for Participation* and T. Kenny Fountain's *Rhetoric in the Flesh: Trained Vision, Technical Expertise, and the Gross Anatomy Lab*, Willerton's book manifests the goal of this series: to develop books that are research driven, interesting, usable, and useful. Willerton's book manifests all four of these characteristics. I think you will find, however, that this book's usability is particularly notable. Most of its body chapters, chapters 4 through 9, end with "key takeaways"—bulleted lists of lessons from the cases that Willerton examines. In addition, this book contains chapter-ending questions and exercises that make it easy to use in pedagogical settings, including TPC courses and industry training.

Willerton's book shows TPC work at its finest. TPC professionals assist people "who need to acquire information and then act on it" (see p. 74). Willerton's BUROC model facilitates this help. With it, TPC professionals can better "identify situations in which plain language can benefit readers" (see p. 73). In these bureaucratic, unfamiliar, rights-oriented, and critical situations, TPC professionals can put their knowledge and skills to work in order to help readers "feel more at ease, understand more about their situations, and make decisions more confidently" (see p. 74). Certainly, for many of us, such important, ethical work played a role in our choice to study and practice TPC.

I am proud to have this book join Potts's book and Fountain's book in the series and, once again, I am looking forward to the other books that will follow.

Jo Mackiewicz
Editor, ATTW Book Series in
Technical and Professional Communication
October 28, 2014

PREFACE

Brief Description of This Project

After I entered the field of technical communication in the 1990s, I supplemented my on-the-job training with resources from the Society for Technical Communication (STC) and classes toward a master's degree in English. One of my classes was about style in technical writing. Through that class, I learned about the Plain Language Action and Information Network (PLAIN), a group of US government employees who provided training and support for federal workers who used plain language to communicate with their organizations' constituents. I enjoyed reading PLAIN's before-and-after examples of documents that writers had revised according to principles of plain language, and I thought the plain-language approach made a lot of sense.

As I continued to work as a technical communicator and to become a scholar in technical and professional communication (TPC), I have been both surprised and a bit dismayed that plain language has not received much attention in the field's publications. At the same time, interest in plain language from people in industry has grown by leaps and bounds—not only in the US, but in countries around the world.

As legislators around the world pass more laws that require communication in plain language that audiences can understand easily, the time is right for an in-depth study of plain language. The practical and financial benefits of using plain language are well documented by groups who advocate for plain language and by authors such as Kimble (2006; 2012). This book, however, is the first to focus on the ethical impacts of plain language: plain language gives citizens and consumers better access to their rights, and it combats the information apartheid that convoluted, overly complicated documents generate.

Audience

I wrote *Plain Language and Ethical Action* for two groups: those who produce plain-language technical content in a variety of fields (online content writers, health literacy specialists, technical communicators, information designers, and content strategists, as well as their managers and supervisors), and those in academic programs such as information design, TPC, and health communication who train these content creators. To serve both audiences well, I combined theoretical views from the academic perspective with practical examples from practitioners of plain-language communication that explain both the *how* and *why* behind their choices. The book's focus on ethics affirms our field's connection to humanistic ideals, and it gives practitioners another means to understand the impacts of their work.

The book could serve as a text for capstone courses in technical communication, graduate seminars, ethics courses, and editing courses. Outside of academia, practitioners can use this book in their professional development.

Structure of the Book

The book has three parts. The first part begins with an overview of the worldwide plain-language movement and introduces the BUROC (bureaucratic, unfamiliar, rights oriented, and critical) model of situations in which plain language can support ethical action. Next, a review of the literature on ethics in technical communication explains how Buber's view of dialogic ethics—already recognized in the TPC literature—provides an approach for understanding plain language. The section concludes with insights on the ethics of plain language from my interviews with plain-language advocates around the world. The second part offers six profiles of organizations using a dialogic approach to create plain-language content for audiences facing challenging BUROC situations. These organizations range from volunteers in an online, grassroots advocacy effort to a nonprofit company with hundreds of employees. The profiles provide insights on how and when to use plain language for a range of organizations and audiences. The third part examines innovative applications of plain language and concludes with principles for applying plain language that derive from the six profiles.

Despite its increasing use around the world, plain language does not appear prominently in the TPC literature; perhaps this results from the field's traditional focus on user documentation in industry. In the twenty-first century, however, many new jobs involve creating content in areas such as health literacy and government services where many advocate for plain language. At the same time, those who teach current and future generations of content creators need to know what plain language is and how it impacts audiences. This timely book examines both why and how writers might create technical content in plain language.

Distinctive Features of This Book

One prominent feature of this book is its model for identifying situations in which plain language supports ethical action. Although I will not argue that every decision to use plain language is a decision involving ethics, BUROC situations are important to view through the lens of ethics because of their strong connections to individuals' rights; any situation involving human rights involves ethics. The BUROC model identifies opportunities for ethical action in situations that technical communicators face frequently.

- B is for *bureaucratic*. These situations involve bureaucracies with policies and procedures that must be followed assiduously. Often the decision makers with whom people need to communicate are in a distant location or are hidden behind the bureaucracy's public façade.
- U is for *unfamiliar*. Citizens might face such situations rarely or at least infrequently. Jargon, policies, and even facilities that citizens must use are not immediately at their command or recollection.
- R and O are for *rights oriented*. These situations affect individuals' choices to act within their rights—rights as citizens, as patients, as consumers, as humans.
- C is for *critical*. These situations are weighty, they are important, and they should not be regarded lightly. They can have significant consequences for people facing them.

Other books on ethics in TPC provide extensive analysis of ethical scenarios that are drastic and dramatic, such as stealing intellectual property, fabricating or misrepresenting data, or whistleblowing. Both the BUROC model and the book as a whole show that ethical issues in TPC are not limited to these situations that technical communicators are likely to encounter infrequently.

Another prominent feature of this book is that it examines the intersection of plain language and ethics in depth. Because theorists identify the field of technical communication as rhetorical and humanistic—for example, see Miller (1979) and Ornatowski (1992), among many others—it is important to continually assess and examine how technical communicators face and respond to ethical situations. Previous discussions of plain language and ethics, as in Brockman (1989b) and Graves and Graves (2011), have been only cursory. In an ethics anthology for the STC, Brockmann (1989b) writes that plain language is not an ethical concern. On the other hand, Graves and Graves (2011) assert that technical communicators have ethical responsibilities to their audiences, and yet they do not discuss those responsibilities in detail. Through the BUROC model and the literature on ethics in technical communication, we can see that plain language empowers consumers and citizens whose rights are often hidden from them by complex, bureaucratic language. Supporters of plain language have long argued that plain language is effective for reaching audiences. For example, Kimble's (2012) latest book identifies 50 projects around the world in which plain language proved expedient and useful. But as Katz (1992) shows, focusing only on what is expedient may—in extreme cases—lead workers or organizations to lose sight of humanistic values entirely. This book makes a new contribution to the scholarship on technical communication by identifying the ethical impacts of plain language.

A third feature of this book is that each chapter includes questions and exercises on important concepts. These questions and exercises will benefit both readers who are students and readers who are not as they reflect on each chapter.

Descriptions of the Chapters

Chapter 1 sets the stage for understanding the intersection of plain language and ethics. I begin by discussing conflicting perspectives on ethics and plain language in technical communication. Next, I provide an overview of the worldwide movement toward plain language. The chapter identifies key points in plain-language history around the world over several centuries. I identify major initiatives and organizations supporting plain language, and I acknowledge common criticisms of plain language. I then introduce the BUROC model for situations in which plain language supports ethical action. I close the chapter with a preview of subsequent chapters.

Chapter 2 reviews the literature on ethics in TPC. I begin by identifying professional organizations providing publications on professional topics, including ethics. Next, I discuss foundational principles of ethics in technical communication; these include Kantian views of imperative behavior, utility, and feminist approaches including care. The chapter continues with a review of the field's articles and essays on ethics, which fit primarily into professional and academic categories. I offer Buber's dialogic ethics as a model for understanding ethical implications of using plain language.

Chapter 3 connects theory with practice by presenting views on ethics from leading plain-language practitioners around the world collected through interviews that I conducted. This chapter includes the professionals' definitions of ethical behavior for plain-language writers; their views on whether "Plain language is a civil right," as Vice President Al Gore once notably stated; their views of the strengths and weaknesses of Buber's I–You dialogic ethics model; and their insights into the strengths and weaknesses of the BUROC model of opportunities for ethical plain-language use.

In chapter 4, I give a profile of plain-language work at Healthwise, a nonprofit company providing health information in several formats for many audiences. Many BUROC situations that consumers face are related to health and medical issues. The chapter gives a background of the organization and states the connection between Healthwise content and BUROC situations. I describe the organizational culture at Healthwise and show how its commitment to plain-language content permeates the company. I discuss the practices Healthwise employees use to ensure their content is appropriately plain for their audiences. I conclude with lessons others can learn from the approach of Healthwise.

Chapter 5 describes plain-language content creation at Civic Design, focusing on the Field Guides to Ensuring Voter Intent it distributes throughout the US. Civic Design has been an ongoing project more than a company. Dana Chisnell leads the project and brings in other professionals as needed. Voting is one BUROC situation among many relating to citizenship in a democracy, and local election officials face many challenges in overseeing voting activities. In this chapter, I describe how and why Civic Design creates content in plain language. I identify practices that support dialogue between Civic Design and audience members, and I provide lessons that others may take away from the approach of Civic Design.

Chapter 6 provides a profile of the plain-language work to revise or "restyle" the Federal Rules of Evidence, which govern how evidence may be introduced into US courts. A series of committees worked together on the project over five years. This chapter cites the extensive records of the committee's work and includes interviews with two people who participated significantly in the restyling work: Professor Joseph Kimble of Cooley Law School and US District Judge Robert A. Hinkle. The Evidence Rules address BUROC situations that occur in courtrooms every day; decisions about evidence are especially urgent and critical because judges and attorneys must make them quickly, with little advance notice. Many in the legal community oppose plain legal language and prefer the status quo because any change to the rules affects many, many people. In this chapter, I describe how proponents of plain language addressed concerns of both critics of change and skeptics of plain language. This chapter provides valuable lessons that others can take away from the approach that committee members followed.

Chapter 7 is a profile of CommonTerms, an international grassroots project to which members contribute through distributed online work and occasional in-person meetings. CommonTerms provides an online tool that generates plain-language summaries of complex terms-and-conditions documents that internet and software users frequently encounter but rarely read. Pär Lannerö leads the project and coordinates work with other professionals. Online privacy is a BUROC issue that worries many; consumers sometimes regret agreements they consent to in their haste to access a service or download a program. I describe how CommonTerms developed and tested its online tool before release, and I identify lessons others can take away from CommonTerms.

Chapter 8 is a profile of the work of Health Literacy Missouri (HLM), a nonprofit organization based in St. Louis, Missouri. Citizens with low health literacy regularly face BUROC situations affecting their health and well-being, and experts estimate that one-quarter of Missourians struggle with low health literacy. HLM integrates plain language and health literacy in innovative ways that empower clients to improve their own communication skills. HLM focuses on training health-care providers, reviewing clients' documents to improve their clarity, and raising awareness about health literacy. I describe how HLM content and services address BUROC situations, who creates plain-language content at HLM, and how ethics and organizational culture affect HLM's work. The chapter closes with lessons that plain-language professionals can take away from HLM.

Chapter 9 provides a profile of the work done by a small consulting firm, Kleimann Communication Group, to redesign mortgage disclosure documents required by the Truth in Lending Act (TILA) and the Real Estate Settlement Procedures Act (RESPA). Working with the recently formed Consumer Financial Protection Bureau (CFPB), Kleimann Communication Group used iterative development and frequent usability testing to create two industry-standard documents that are required for mortgage-loan transactions covered by TILA and RESPA. Kleimann Communication Group worked with home-purchase professionals and consumers in nine cities across the US to obtain a significant amount of feedback on proposed designs. This chapter provides valuable lessons that others can take away from the approach that Kleimann Communication Group followed. Questions and exercises follow.

In chapter 10, I discuss prominent applications of plain language to technical content in public settings. Although the people developing these applications have not explicitly aligned themselves with advocates for plain language, they do create clear, accessible content while gathering feedback through dialogue with their audiences. These content creators include Mignon Fogarty, creator of Grammar Girl and the Quick & Dirty Tips websites and books; John Wiley & Sons, producer of the For Dummies series; and Common Craft, a Seattle duo who pioneered the development of online explanation videos. While some of these content creators address BUROC situations more than others do, collectively they provide content that empowers consumers to act and challenge the power differential that separates experts from nonexperts. Chapter 11 concludes the book by summarizing major insights from previous chapters and by asserting that plain language can support ethical action. I return to the BUROC framework and assess its value as a heuristic for identifying difficult situations and reflecting on how plain language can benefit the individuals facing them.

ACKNOWLEDGMENTS

I am grateful for the valuable time I received to focus on this book. I thank Boise State University—especially President Robert Kustra and Provost Martin Schimpf—for providing a semester of sabbatical research leave, and I thank the Boise State University Arts and Humanities Institute (AHI) for providing a semester's leave as an institute fellow and funds for technology and travel. I also thank AHI Director Nick Miller and Vice President of Research and Economic Development Mark Rudin for their support. Thanks to Albertsons Library at Boise State University for providing access to many resources, and thanks to librarians Carrie Moore and Memo Cordova for assistance. I thank David Monroe and Patrick Robinson for technical support.

For support of this project in its early stages, I thank Amy Koerber, Karen Schriver, Roger Munger, Michelle Payne, Clay Morgan, and Tony Roark. Fruitful conversations with Mike Markel, Whitney Quesenbery, and Jacky O'Connor helped me greatly. Christopher Trudeau, Cheryl Stephens, Deborah Bosley, and Joseph Kimble connected me to several plain-language professionals and provided essential encouragement. Derek Ross, Laura Palmer, Jonathan Arnett, Carlos Evia, Dave Yeats, and Sam Dragga provided valuable comments on early chapter drafts. Kim Sydow Campbell suggested placing a moral compass for plain language on this book's cover.

I am grateful to all the professionals who discussed their plain-language work with me. To each person in chapters 3 through 10 who made time to speak or correspond with me about working with plain language, I give my sincere thanks. Several professionals in chapter 3 provided comments on a draft of the chapter, and I thank them.

I thank Healthwise, Inc., for giving me so many opportunities to conduct interviews and observe its work. Thanks to Healthwise CEO Don Kemper for showing me Elspeth Murray's poem that I mentioned in chapter 1, to Karen

Baker for supporting this project and giving her employees time to speak with me, and to Christian Lybrand for coordinating reviews of chapter 4.

I have sincerely enjoyed working with the staff of Routledge. I'm especially grateful to Ross Wagenhofer for his help through the production process and to Linda Bathgate for her continuous support of this project. As editor of the ATTW Book Series, Jo Mackiewicz worked diligently through multiple drafts to help me transform my ideas for a project into this book. I am grateful for her patience, editorial skill, and guidance throughout a long process. Thanks to the entire ATTW Book Series editorial team for their outstanding work and to Lori Peterson for expert copyediting. Thanks to Ross, Jo, and the Routledge team for help creating the moral compass for plain language on the book's cover. Thanks also to the external reviewers for their work: Sam Dragga, Janice (Ginny) Redish, and Michael Salvo.

Finally, I thank my family, Cecily and Owen, for supporting me throughout this project in every way.

Russell Willerton
December 2014

1

UNDERSTANDING PLAIN LANGUAGE AND OPPORTUNITIES TO USE IT

Over the past several decades, advocates around the world have urged people writing for audiences of consumers and citizens to use plain language. According to UK plain-language advocate Martin Cutts (2009), plain language is the "writing and setting out of essential information in a way that gives a cooperative, motivated person a good chance of understanding it at first reading, and in the same sense that the writer meant it to be understood" (xi). Steinberg (1991a) writes that plain language "reflects the interests and needs of the reader and the consumer rather than the legal, bureaucratic, or technological interests of the writer or the organization the writer represents" (7). Advocates for plain language around the world have identified principles of word choice, verb selection, sentence construction, visual design, organization, and usability testing that make complex technical documents easier for nonexpert consumers and citizens to use (Cutts 2009; Steinberg 1991b). When constituents can use their documents more quickly and effectively, companies and government agencies save money; for example, Kimble (2012) provides 50 examples of plain-language projects that saved readers time and money and that reached audiences effectively. While people tend to use the terms "plain language," "plain English," and "plain writing" interchangeably (Greer 2012), "plain language" is the most inclusive of these terms. Plain-language practitioners around the world apply key concepts from the movement usefully in many languages and cultures.

Although plain language has grown in prevalence around the world, researchers have done little work to understand the degree to which plain language is a means for technical communicators to do ethical work. Because theorists believe that the broad field of technical communication is rhetorical and humanistic—for example, see Miller (1979) and Ornatowski (1992), among many others—it is important to continually assess and examine how technical communicators face and respond to ethical situations. Two examples show how some disagree over whether plain language is ethical. The first is from Brockmann's (1989b) introduction to *Technical Communication and Ethics*, a collection of articles and essays published by the Society for Technical Communication. Brockmann explains why the collection does not address plain language: "Plain language, although a readability concern, is not necessarily an ethical concern. Identification of plain language with ethical language mistakes the outward signs of ethics, plain language, for true ethical actions" (v). Writing more than two decades after Brockmann, Graves and Graves (2011) state in their textbook on technical communication that "at its heart, plain language involves an ethical relationship between the reader and writer. As a writer, you must want to communicate with your audience clearly" (71). In this book, I explore the extent to which a middle ground exists between these two examples while reaffirming the humanistic concerns of technical communication.

Over the past several decades, plain-language advocates around the world have worked for clearer government forms, laws that citizens can readily understand, and letters that clearly explain how to obtain government benefits. The US now has its first federal law requiring plain-language activities in government agencies, the Plain Writing Act of 2010. Perhaps this is an opportune time to determine whether a new perspective on plain language and ethics exists between the perspectives that Brockmann and Graves and Graves articulate. To that end, this book focuses on two main questions:

- Is plain language an ethical concern?
- What processes and procedures can help plain-language writers do ethical work that helps their audiences?

An Overview of the Worldwide Movement toward Plain Language

Several authors and editors provide insight into the development of the plain-language movement around the world. Concern for plain language has often focused on documents produced by government agencies, but it now extends to law, health and medicine, and many aspects of business. A review of the plain-language movement's history helps identify the social forces behind the movement and demonstrates that concerns about confusing, bureaucratic language are long-standing.

Early Developments in Plain-English Style

References to plain-English style date back to the fourteenth century. See table 1.1 for a brief list of developments in plain-English style between the fourteenth and seventeenth centuries.

TABLE 1.1 Early developments in plain-English style.

Century	Development in plain-English style
Fourteenth Century	The Host in Chaucer's *Canterbury Tales* exhorts the learned Clerke of Oxenford to speak plainly so the pilgrims may understand him (McArthur 1991, 14).
Sixteenth Century	Writers of technical books in English in the sixteenth century used plain style for their audiences. But because writers used plain-English style outside of traditional literary genres, this style choice did not receive much attention (Tebeaux 1997).
Seventeenth Century	The first person to refer to "plain English" as opposed to florid, ornate English may be Robert Cawdrey, who compiled the first known English dictionary in 1604. His *Table Alphabeticall* uses English words to define difficult words borrowed from languages like Latin, Greek, and Hebrew. Cawdrey's stated audience for the dictionary was women. Without formal education, women had "no easy way of appreciating the layer of Latinity that had formed, as it were, along the top of traditional English" (McArthur 1991, 13).
	Francis Bacon prominently advocated for plain style in science, as did the Royal Society. Women writers such as Margaret Cavendish and Jane Sharp used plain style to reach their audiences (Tillery 2005).

Influences on the Plain-Language Movement in the Early- and Mid-Twentieth Century

As table 1.2 shows, activities in the US and the UK influenced developments in plain language in the early- and mid-twentieth century.

New Momentum in the 1970s

According to Redish (1985), before 1970 bureaucrats faced no mandates to write in ways that consumers could understand. In the 1970s, presidential executive orders and changes in federal and state laws helped give legitimacy to the plain-language movement. Fervent consumer activism and an increase in government paperwork brought new attention to the problems of unclear bureaucratic language. Table 1.3 lists several developments from this decade in the US and the UK.

Responses in the US to a Particular Type of Unclear Language

In the early 1970s, the National Council of Teachers of English (NCTE) took a stand against a particular type of unclear language. Advertisers, politicians, and others in the media during that era intentionally used unclear language, called "doublespeak" in reference to George Orwell's novel *1984*, to mislead public

TABLE 1.2 Influences on the plain-language movement from the 1900s to the 1970s.

Area of activity	Influence on the plain-language movement
US Government	Maury Maverick, once chairman of the Smaller War Plants Corporation, wrote a memo in 1944 to everyone in the corporation requesting that lengthy memoranda and "gobbledygook" language be replaced by short and clear memoranda. Maverick coined the term "gobbledygook" after recalling the sights and sounds of a bearded turkey strutting and gobble-gobbling about (Greer 2012, 4).
	In 1953, Stuart Chase wrote *The Power of Words*, which includes a chapter bemoaning gobbledygook in bureaucracies, law, and universities. In 1966, John O'Hayre of the Bureau of Land Management released 16 essays on plain-English writing for business and government in *Gobbledygook Has Gotta Go* (Redish 1985, 128).
UK Government	Ernest Gowers advocated for civil servants to communicate clearly. A training pamphlet he wrote in 1943 later grew into the book *Plain Words* in 1948. Gowers published a companion reference book, *The ABC of Plain Words*, in 1951. Gowers combined those two books into *The Complete Plain Words* in 1954, a volume reprinted many times (Kimble 2012, 51–52).
	In 1946, British author George Orwell complained about the "slovenliness" of writing about government and politics in modern English. In his classic essay, "Politics and the English Language," which has appeared in writing anthologies for decades, Orwell (2005) provides six succinct rules to help writers remove vagueness and pomposity.
US Education and Research	Although studies of factors affecting the readability of texts date back as far as the 1890s in the US, research on readability increased notably after researchers surveyed and tested adult literacy. The military started literacy surveys in 1917; other agencies began testing civilians and students soon after (DuBay 2004).
	Prominent researchers from this era include Rudolph Flesch, who released his Reading Ease formula in 1948 (Kimble 2012, 49–50), as well as William S. Gray and Bernice Leary, Irving Lorge, Edgar Dale and Jeanne Chall, Robert Gunning, Wilson Taylor, and George Klare (DuBay 2004).
	Researchers have developed hundreds of readability formulas over time. Experts have long debated how and whether these formulas should be used, especially because the formulas focus on a text's surface features—numbers of syllables, words, and sentences. Readability formulas are important in the history of plain language because they have influenced understandings of plainness and clarity and because some laws and regulations require plain-language texts to meet particular readability scores.

TABLE 1.3 Developments in the plain-language movement in the 1970s.

Government entity	Development in the plain-language movement
US Government	In 1972, President Richard Nixon decreed that the *Federal Register* should use layman's terms and clear language (Dorney 1988).
	In 1977, the Commission on Paperwork issued a report strongly recommending that the government rewrite documents into language and formats that consumers could understand (Redish 1985, 129).
	Congress also passed several laws that required warranties, leases, and banking transfers to be clear and readable. These included the Magnuson-Moss Warranty-Federal Trade Commission Act of 1973, the Consumer Leasing Act of 1976, and the Electronic Fund Transfer Act of 1978 (Greer 2012, 5).
	President Jimmy Carter issued executive orders requiring plain language in 1978 and 1979. Executive Order 12044 set up a regulatory reform program that required major regulations to be written in plain English so that constituents could comply with them. Executive Order 12174 required agencies to use only necessary forms, to make the forms as short and simple as possible, and to budget the time required to process paperwork annually. President Carter also signed the Paperwork Reduction Act, which took effect after he left office (Redish 1985, 129).
	President Ronald Reagan rescinded Carter's orders requiring plain language. Reagan did, however, support regulatory reform and the Paperwork Reduction Act (Redish 1985, 129–30). Reagan's secretary of commerce, Malcolm Baldridge, argued for using plain language in business and industry (Bowen, Duffy, and Steinberg 1991).
State of New York	In 1975, Citibank shocked the financial community by dramatically simplifying a loan document. The original loan note had about 3,000 words, but the revised note had 600 (Redish 1985, 130).
	In 1977, New York became the first state to enact a law requiring plain language. Named the Sullivan Law for its sponsor, Assemblyman Peter Sullivan, it requires businesses (including individual landlords) to write contracts with consumers using words with common, everyday meanings (Felsenfeld 1991).
	At least ten other states later followed New York's lead. Critics predicted a wave of lawsuits over the new contracts, but few filed lawsuits (Kimble 2012, 54–56).

(Continued)

TABLE 1.3 (Continued)

Government entity	Development in the plain-language movement
UK Government	The 1974 Consumer Credit Act became the first British law to require plain English. It requires credit-reference agencies to give consumers, upon request, the contents of their files in plain English they can readily understand (Cutts 2009, xvii).
	In a 1979 protest in Parliament Square, campaigners for plain English publicly shredded unclear government forms. The event helped persuade the incoming Margaret Thatcher administration to issue new policy about government forms. Agencies had to count their forms, remove unnecessary forms, revise the rest for clarity, and report their progress to the prime minister annually. Many local governments also followed suit (Cutts 2009, xv–xvi).

TABLE 1.4 Responses to doublespeak in the US.

NCTE formed the Committee on Public Doublespeak to educate students and teachers about the dangers of doublespeak and to expose those who manipulated language. The committee published the *Public Doublespeak Newsletter*, which eventually became the *Quarterly Doublespeak Review* (National Council of Teachers of English 2009).

Amid such lying and deliberate obfuscation, Perica (1972) argued that the Society for Technical Communication (STC) needed a strong code of ethics.

In the aftermath of another 1970s scandal, the Watergate hotel break-in and cover-up that cost Richard Nixon the US presidency, STC drafted a code of ethics and offered it to the membership (Malone 2011).

audiences. While many bureaucratic documents embody sloppy thinking, convoluted vocabulary, and poor efforts at communication, writers carefully craft doublespeak to mislead audiences and distort reality (Lutz 1988, 41). Table 1.4 provides a brief list of responses to doublespeak in the US.

Important Research on Plain Language and Document Design

Schriver (1997) discusses important collaborations between experts from industry and academia that led to research and practical knowledge about how readers understand plain-language documents. These studies brought attention to documents that, while ubiquitous, researchers had rarely analyzed systematically. Table 1.5 lists some milestones in research on plain language and document design.

TABLE 1.5 Milestones in research on plain language and document design.

The National Institute of Education funded the Document Design Project (DDP) between 1978 and 1981. The American Institutes for Research (AIR), certain Carnegie Mellon University faculty, and the firm Siegel & Gale joined together in the project. DDP provided training to personnel in many federal agencies.

DDP produced two books that strongly influenced the field of document design: *Document Design: A Review of the Relevant Research* (Felker 1980) and *Guidelines for Document Designers* (Felker et al. 1981).

After 1981, research continued at AIR's Document Design Center and at Carnegie Mellon University's interdisciplinary Communication Design Center. AIR's Document Design Center became the Information Design Center in 1993. Although the Communication Design Center (CDC) was by all accounts successful, it closed after 1990 in the wake of changes among Carnegie Mellon administrators.

Interest in Plain Language in the Legal Community

Kimble (2012) identifies important publications that promote the use of plain language by lawyers, judges, and law professors. Table 1.6 lists the books and journals prominently promoting plain legal language.

TABLE 1.6 Publications promoting interest in plain language among the legal community.

Year	Book
1963	David Mellinkoff's *The Language of the Law* gave scholarly weight and "undeniable validity" to criticisms of legal writing going back for centuries; it provided "the intellectual foundation for the plain-language movement in law" (Kimble 2012, 47).
1979	Richard Wydick's *Plain English for Lawyers* provided concrete advice on removing surplus words, choosing familiar words, and crafting effective sentences. More than 800,000 copies of five editions have sold over 30 years (Kimble 2012, 48).
1984	*Michigan Bar Journal* first produced its column on plain language. Kimble called it "the longest-running legal-writing column anywhere" (Kimble 2012, 49).
1990	Australian lawyer Michèle Asprey first released *Plain Language for Lawyers.* Kimble calls it—now in its fourth edition—"the single most comprehensive book on the subject" and says it has influenced lawyers around the world (2012, 54).
2001	Bryan A. Garner first published *Legal Writing in Plain English*, a text with exercises for legal professionals.

Plain-Language Movement in the US in the 1990s

In the 1990s, the plain-language movement gathered significant momentum in the US. Table 1.7 summarizes major events.

TABLE 1.7 Developments in the US plain-language movement in the 1990s.

Years	Development in the US plain-language movement
Mid-1990s	Federal employees in the Washington, DC, area began meeting to discuss plain-language issues. Originally called the Plain English Network, the group is now the Plain Language Action and Information Network (PLAIN). PLAIN hosts regular meetings and offers training on plain-language writing to federal agencies (Plain Language Action and Information Network 2013).
Mid-1990s	The Securities and Exchange Commission (SEC) offered a shorter review period to corporate volunteers willing to file plain-language disclosure documents. In September 1996, Bell Atlantic and NYNEX, which were planning to merge, mailed what was probably the first joint proxy statement written in plain language. During that project, the SEC also drafted its *Plain English Handbook*, which is publically available. In 1998, the SEC adopted rules requiring companies to write investment prospectuses in plain English (Kimble 2012, 56).
Late 1990s	President Clinton revived plain language as a major government initiative. Clinton issued a memorandum that formalized the requirement for federal employees to write in plain language (Locke 2004). Clinton's memo directed leaders of executive departments and agencies to use plain language in all new documents (other than regulations) that explain how to obtain a benefit or service or how to comply with a requirement they administer or enforce. Vice President Al Gore monitored this initiative.
	Vice President Gore presented monthly No Gobbledygook awards to federal employees who turned bureaucratic messages into plain language for citizens. The iconic statement "Plain language is a civil right" came from Gore (Dieterich, Bowman, and Pogell 2006).

Progress for Plain Language in Other Countries around the World

The US was not the only country supporting a burgeoning plain-language movement. Table 1.8 summarizes developments in countries around the world.

Recent Developments in the US

Since 2000, major US government agencies have significantly increased their commitments to using plain language. The Federal Aviation Administration, the National Institutes of Health, and the Department of Agriculture are among those agencies with the strongest plain-language programs (Locke 2004).

More recently, US lawmakers wrote plain language into federal law. The Plain Writing Act of 2010, which President Barack Obama signed into law, requires federal agencies to demonstrate awareness of plain language, offer plain-language training, and write new public documents in plain language. Agencies must also

TABLE 1.8 Developments in the plain-language movement around the world.

Country	Development in the plain-language movement
Australia	In the 1970s, Australia featured the first plain-language car insurance policy. In 1984, the government adopted plain-language policy for its public documents; this policy now extends to the language of the law itself (Cutts 2009, xxi).
	In the early 1990s, two influential reports shaped the content and design of Australian legislation (Kimble 2012, 75). From 1990 to 1996, the grant-funded Centre for Plain Legal Language at Sydney University published regular columns and conducted research on the economic benefits of plain-English documents (Kimble 2012, 99).
Canada	The Alberta Law Reform Institute has encouraged plain language in the law since 1968 (Kimble 2012, 85).
	From 1973 to 1992, the Canadian Legal Information Centre worked to improve the legal information and public legal literacy. A national nonprofit coalition, it created a Plain Language Centre to promote legal documents in plain English and plain French (83–84).
	The Plain Language Service at the Canadian Public Health Association (CPHA) offers plain-language revisions of health materials and training on clear communication. Once funded by the federal government, the Plain Language Service is now an innovative, self-financed part of the CPHA (86–87).
European Union	A 1993 EU directive requires businesses to write consumer contracts in plain language and to negotiate them in good faith. Many member countries have written this directive into their own national legislation (Kimble 2012, 59).
	In 1998, the European Commission, which runs the EU, started the Fight the Fog campaign to promote plain language. This effort relaunched in 2010 as the Clear Writing Campaign. It offers a booklet, *How to Write Clearly*, in all 23 official languages of the EU (Kimble 2012, 91–92).
	The European Commission is developing an interdisciplinary training course in clear communication through the IC Clear consortium. The course will combine training in plain language, information design, and usability (International Consortium for Clear Communication 2011).
New Zealand	The Law Commission has produced influential reports on making legislation more accessible and understandable. The Parliamentary Counsel Office has adopted many Law Commission recommendations in New Zealand statutes and regulations (Kimble 2012, 100–104).

(Continued)

TABLE 1.8 (Continued)

Country	Development in the plain-language movement
Nordic Countries	Advocates in Sweden have influenced language in legislation; a group of reviewers within the ministry of justice must vet bills before they can be printed. A government-sponsored group, Klarspråksgruppen, encourages agencies to write in plain language (Cutts 2009, xxii).
	Similar projects in Denmark, Finland, Iceland, and Norway encourage government agencies and officers to write public documents in clear language that constituents can understand (Kimble 2012, 92–95).
	Sweden's Stockholm University is probably the first in the world to offer a degree focused on plain language. Graduates from the Swedish Language Consultancy work in public and private sectors (Kimble 2012, 97).
South Africa	The South African Constitution (Republic of South Africa 2014) is a crucial government document written in plain language. Many parties developed the principles of the Constitution through dialogue before South Africa's first democratic elections (Kimble 2012, 60–61).
	South Africa's Consumer Protection Act of 2008 requires business and agencies to give information to consumers—notices, documents, and visual representations—in plain language that an ordinary consumer can understand. Penalties for failing to comply are substantial (Kimble 2012, 62–63).

post annual reports on their compliance with the Plain Writing Act on their websites (Plain Language Action and Information Network 2013). Critics note that the Plain Writing Act is neither subject to judicial review nor enforceable by administrative or judicial action, nor does it address federal regulations.

Early drafts of the Plain Writing Act contained a provision requiring federal agencies to write regulations in plain language, but supporters dropped that provision after some legislators opposed it. Iowa congressman Bruce Braley, who introduced the Plain Writing Act, has authored the Plain Regulations Act to require federal regulations in plain language (Cheek 2012). Braley introduced the bill in both the 112th and 113th sessions of Congress, but it did not advance out of committee. Because federal regulations affect many citizens, especially small business owners, advocates hope the bill will continue through the legislative process.

Organizations Promoting Plain Language

Around the world, many organizations advocate for plain language in information provided to citizens and consumers. Some organizations and initiatives are listed in tables 1.7 and 1.8. In the US, the Center for Plain Language (CPL) advocates for plain-language laws and regulations. (It was actively involved in the effort to pass the Plain Writing Act.) CPL also provides training and education and gives annual awards. CPL's ClearMark awards go to the best examples of plain language from government, nonprofits, and private businesses while its WonderMark awards go to poorly written documents. CPL also provides an annual report card to grade federal agencies on the quality of their public communication and their compliance with the Plain Writing Act. Clarity International is an international group advocating for plain language in legal documents. Clarity hosts a biennial conference at sites around the world, and it publishes a journal, *Clarity*, twice each year. Plain Language Association International (also known as PLAIN but different from the Plain Language Action and Information Network in the US) is another international organization promoting plain language in all areas of business and government. Formed in 1993 as the Plain Language Consultants Network, PLAIN hosts conferences around the world every two to three years.

Resistance against Plain Language

While the advocates of plain language are quick to identify its benefits, some believe the approach has shortcomings. Although no organizations or formal coalitions campaign against plain language, critics have documented their concerns. Both Mazur (2000) and Kimble (2012) compile and respond to common concerns about plain language such as these:

- Plain language is a concept too broad to be useful.
- Plain language involves following rules slavishly.
- Plain language is only about shortening texts and dumbing them down.
- Plain language means writers cannot use technical vocabulary.
- Plain language relies on readability formulas that have questionable validity.
- Plain language is not as precise as typical bureaucratic or legal language.
- Readers of legal and bureaucratic documents do not like or want plain language.

Mazur (2000) and Kimble (2012) address each complaint in their respective works. Some complaints, such as "plain language involves following rules slavishly," quickly lose merit after a review of the plain-language literature. Others, such as "plain language is not as precise as typical bureaucratic or legal language," often come from people who believe changes to the status quo will threaten their sources of income; as a professor of law and legal writing, Kimble has spent a career challenging such entrenched positions.

Some have raised concerns that plain English might present problems to nonnative English speakers. For example, many plain-language guidelines suggest selecting simple English words (often with Germanic roots) over more complex Latinate English words. Maylath (1997a) writes that a nonnative English speaker whose primary language is among the Romance languages might understand Latinate English words more easily than shorter words with Germanic roots. In a pair of studies, Thrush (2001) found that nonnative English speakers may struggle more to understand plain phrasal verbs than Latinate English verbs and that native French and German speakers prefer English words with Latinate roots. In addition, choosing a plain style might not be effective for communicating with other cultures. Maylath and Thrush (2000) write that some cultures prefer nuanced, layered messages over direct messages. They also state that some cultures might regard a writing style that is too short, direct, and simple as a sign of incompetence and lack of sophistication (239–40). Similarly, Riley and Mackiewicz (2003) write that nonnative English speakers might violate expectations for politeness by using plain language in some rhetorical situations.

And yet, Maylath (1997b) points out that clarity, a hallmark of effective plain-language documents, can be a virtue for documents intended for translation: "The easiest texts to translate avoid ambiguity and confusion" and show effective organization (344). A survey of technical translators found that they struggle to collaborate successfully with technical writers who request translations of documents that are not plain, well structured, and written from the user's perspective (Gnecchi et al. 2011, 177).

Just as practitioners define plain language in many ways, they apply its principles differently in response to specific rhetorical situations. Each situation presents its own challenges. Writers cannot guarantee that an audience will understand any text—whether written in plain language or not; myriad factors affect readers' understanding (Cutts 2009, xii). While plain language principles do not provide a panacea for every challenging communication problem, they nevertheless provide ways to help conscientious communicators actively pursue the goal of reaching many intended readers effectively (Kimble 2012, 31–35).

A Model to Identify Opportunities for Ethical Use of Plain Language

The next two narratives exemplify some of the situations in which writers and organizations could use plain language to promote the self-interest of others who face stressful, challenging situations.

Example 1: A Patient's Confusion

Nadia Ali once took her husband to the emergency room at 2:00 a.m. Her husband had severe shoulder pain, nausea, and light-headedness (Ali 2012). The brusque triage nurse directed them down the hall but did not specify a room. The nurse ordered her husband to put on a hospital gown but left before Ali could say that his shoulder pain would make that difficult. Around 3:00 a.m., her husband finally saw a physician. Ali, who is a hospitalist physician herself, described the remainder of their visit:

> My husband, who is educated and intelligent, got more confused with all the jargon used by the physician. I asked him why he didn't ask any questions. He said he wasn't sure it would be helpful in case the doctor repeated the same technical terms again. He also felt the doctor was possibly in a rush since he stood the whole time. Finally, around 4:00 a.m., my husband got some pain and nausea medication. We left the ER around 6:00 a.m. with a script [prescription for medicine]. Our next task was to find a pharmacy that would be open, since that information was not provided. (2012, para. 5)

Going through this experience with her husband, Ali (2012) realized "the pain and agony [patients] have to go through to get the care they need. Patients and families feel lost and helpless amongst a sea of health care professionals they encounter. . . . Lots of questions and concerns are never addressed." If the physician had used more familiar vocabulary in speaking with Ali's husband, if the hospital staff had provided a list of 24-hour pharmacies to patients with prescriptions to fill at odd hours, and if the triage nurse had given even a specific room number—each an opportunity for plain language—Ali's husband might have avoided some of the anxiety he experienced in the hospital.

Example 2: Voters' Uncertainty

The second example of an opportunity to benefit others through using plain language comes from the state of Kansas. In February 2012, voters in the city of Wichita considered a referendum on whether the developer of the Ambassador Hotel downtown should get to keep a portion of the hotel's future guest-tax revenue (Lefler 2013). A "yes" vote supported giving the developer the tax break while a "no" opposed it. Unfortunately, the complicated wording of the referendum left many voters confused. The language on the ballot question read as follows:

> Shall Charter Ordinance 216 entitled: "A charter ordinance amending and repealing Section 1 of Charter Ordinance No. 213, of the city of Wichita, Kansas, which amended and repealed Section 1 of Charter Ordinance No. 183 of the city of Wichita which amended and repealed Section 1 of Charter Ordinance No. 174 of the city of Wichita, Kansas, pertaining to the application of revenues from the transient guest tax" take effect? (Lefler 2013)

Election workers received a flood of phone calls and questions about the referendum, but election supervisors instructed them to answer only "Yes means yes.

No means no." The attorney who wrote the language said he had to comply with the requirements of the state constitution and that he could not add explanatory information.

In 2013, the Kansas legislature approved a bill allowing county election officials to request that a designated official write an "explainer" in plain language when the language in a ballot measure is confusing or contains too much legalese for voters to understand easily. Under this policy, a second official reviews each explainer for accuracy and neutrality (Lefler 2013). These explainers help voters exercise their rights to vote and to participate in the democratic process.

Empathizing and Acting through the BUROC Model

Elspeth Murray (2006), a poet in the UK, summarizes the feelings and frustration of someone dealing with a serious medical problem in her aptly titled poem "This Is Bad Enough." Murray complains about poorly written, shoddily reproduced materials that confuse patients instead of helping them. She calls out medical providers who leave patients "adrift" and "lost in another language." She asks for empathy, clear information, helpful visuals, and time to process what a provider says. Murray (2006) concludes the poem poignantly:

> Because this is bad
> and hard
> and tough enough
> so please speak
> like a human
> make it better
> not worse.

Situations like those Ali, Lefler, and Murray describe are just some examples of the situations I identify and analyze with concepts that together form the acronym BUROC (byoo-rok):

- B is for *bureaucratic*. These situations involve bureaucracies with policies and procedures that individuals must follow assiduously. Often the decision makers with whom people need to communicate are in a distant location or are behind the bureaucracy's public façade. For example, buying insurance and making insurance claims are bureaucratic processes.
- U is for *unfamiliar*. People sometimes face situations that are unfamiliar or occur infrequently. Jargon, policies, and even facilities that people must use are not immediately at their command or recollection. For example, an ill patient considering enrolling in a clinical trial likely faces unfamiliar terms and concepts.
- R and O are for *rights oriented*. These situations are especially important because they affect individuals' choices to act within their rights—rights as

citizens, as patients, as consumers, as humans. Instructions on how to obtain an absentee ballot, for example, affect a citizen's opportunity to exercise the right to vote.

- C is for *critical*. These situations are weighty, serious, and important; they can have significant consequences for people facing them. These situations often arise without warning, and they may require people to make important decisions quickly. For example, a policy document for an organ transplantation network addresses matters of life and death for individuals needing new organs; administrators must make decisions about implementing such policies quickly.

I do not argue that every decision to use plain language involves ethics. Nevertheless, BUROC situations are important to view through the lens of ethics because they involve individuals' rights and because they provide opportunities to assist others in need. To assess the validity of this proposed model, I presented it to an international group of plain-language practitioners and asked for their feedback. Chapter 3 contains their comments on the BUROC model.

In This Book

In the chapters ahead, I argue that the choice to use plain language when interacting with constituents facing BUROC situations is both useful and ethical. I profile organizations that use plain language in ethical ways to benefit their constituents. Each profile identifies the processes and procedures used to create effective plain-language materials that affect ethical situations.

Chapter 2 reviews the literature on ethics in technical and professional communication. It identifies dialogic communication ethics as a way to understand the ethical impact of plain language.

Chapter 3 describes the views of prominent plain-language practitioners about plain language's ethical impacts and about applications for the dialogic model of ethics. These practitioners provide a broad perspective on readers' rights to understand the information they receive from governments and companies. This chapter provides a link between theories of ethics and plain-language practices around the world.

Chapter 4 profiles work at Healthwise, a nonprofit company providing health information for consumers in many print and online formats. Links to the BUROC model include the bureaucracy of health care, unfamiliar situations for patients, rights of patients, and the critical and urgent nature of health and medical problems.

Chapter 5 profiles work by the group Civic Design, which created a set of plain-language guides for county election officials to use for the 2012 presidential election. Dana Chisnell is the principal designer on this project. Ties to the BUROC model include the bureaucracies and laws affecting elections, unfamiliar situations for many poll workers, rights of citizens to vote, and the critical importance of getting things right in a short time.

Chapter 6 describes the work to revise or "restyle" into plain language the Federal Rules of Evidence, which govern the introduction of evidence into US courts of law. These rules address BUROC situations that occur in courtrooms every day. Decisions about evidence are especially urgent and critical because judges and attorneys must make them quickly and with little advance notice; all parties in a lawsuit have a right to fair proceedings. Although many in the legal community are skeptical about plain-language legal documents—including some who participated in the restyling effort—attorneys, judges, law professors, and law students benefit from the restyled rules.

Chapter 7 examines an international, grassroots effort called CommonTerms that advocates for plain-language terms of service (TOS)—the rules a person must follow and the conditions a person must accept to use a service, such as an email account or a social network. TOS tend to be long and dense documents, written in legalese and jargon. One study of common TOS indicated that people need 76 workdays each year to read all the privacy policies for software and online services that they use (Wagstaff 2012). CommonTerms is a BUROC project because it deals with bureaucracies of law and policy, unfamiliarity of jargon for consumers, rights to privacy as internet users, and the critical importance of making sound choices quickly.

Chapter 8 contains a profile of plain-language work at Health Literacy Missouri, a nonprofit organization addressing the causes and effects of low health literacy in the state. The organization provides many services to help others use and benefit from plain language. Links to the BUROC model include the bureaucratic nature of health-care systems and insurance coverage, unfamiliar vocabulary and situations for patients with low health literacy, the rights of patients to make decisions, and the critical nature of both low health literacy and health and medical problems.

Chapter 9 profiles work done by a consulting firm, Kleimann Communication Group, to create new mortgage disclosure forms in the US. Links to the BUROC model include the bureaucracy of organizations and regulations that affect the purchase of a home, unfamiliar vocabulary and processes for homebuyers, individuals' rights to choose homes they can afford along with mortgages on the best terms, and the critical consequences of a home purchase.

Chapter 10 identifies and discusses prominent dialogic applications of plain, clear language to technical content in public settings. Creators of these works do not necessarily call themselves practitioners of plain language, but they show how clear communication can reach a wide audience. These applications include the approaches of Common Craft, a Seattle duo of pioneers in the development of online explanation videos, and the Alan Alda Center for Communicating Science, which uses improvisational theater techniques to help scientists, physicians, and other skilled specialists to communicate more effectively with policy makers and the general public. The chapter identifies how these approaches incorporate dialogue with the audience while addressing issues of ethical importance.

Chapter 11 concludes the book by summarizing major insights from previous chapters and providing suggestions for ethically applying plain language to technical content.

Questions and Exercises

1. Search online for a definition of plain language (or plain English, or plain writing) and compare it with Cutts's definition on page 1. How are the definitions similar, and how do they differ? Review both definitions and describe some of the benefits of plain language in around 200 words.

2. Search online to find a guide for communicating in plain language (or plain English, or plain writing).
   - Identify a guideline that you think you follow well in your own writing, and then identify one you do not follow as well. What about these guidelines makes them easy—or difficult—for you to follow, and why do you think that is so? Explain your response in about 150 words.
   - Identify the organization that posted the guide you found. Why do you think this organization promotes plain language? Why might this organization's audiences benefit from plain-language content? Explain your analysis in about 150 words.

3. Search online to investigate how people use plain language in a technical field that interests you (for example, law or medicine). In some instances, you may need to start searching broadly before narrowing your search. Summarize what you learn in about 300 words.

4. Think of a BUROC situation that you have faced—one in which you needed information to make an important decision. Write around 200 words describing the situation and how it turned out for you. If the resources you used or needed are online, review and describe them as well.
   - What was the situation?
   - Which parts of the BUROC framework (bureaucratic, unfamiliar, rights oriented, and critical) were most salient or prominent in the situation?
   - What information did you get, and was it easy to use and understand? Explain.
   - How did you respond to the situation? (How did you act, or what did you decide?)
   - How do you feel now as you reflect on the situation?

2

OVERVIEW OF ETHICS IN THE
TECHNICAL AND PROFESSIONAL
COMMUNICATION LITERATURE

In this chapter, I review discussions of ethics in the technical and professional communication (TPC) scholarship. I identify broad principles of ethics in TPC and develop a framework for understanding the ethical implications of using plain language in technical content. Scholarship in TPC applies to plain-language communication because both involve helping readers find, understand, and act on information. Furthermore, discussions of ethics are robust in TPC but scant in the literature on plain language. That said, plain-language professionals have written about ethics from time to time (e.g., Lauchman 2010; Mackinnon 1997; Mowat 1999; Osborne 2005).

One means of understanding the ethics of plain language is through the lens of rhetoric. In a presentation for the Plain Language Consultants Network, which later became Plain Language Association International (PLAIN), Mackinnon (1997) said that contemporary views of rhetoric indicate that language is "socially bound and binding, as part of a political and ethical context" (para. 9). Mackinnon notes that the windowpane theory of language, a theory that considers language a neutral tool for transferring meaning, fails to consider how readers and listeners must actively interpret the information they receive. Because language—even plain language—interconnects with social ideas of power, influence, and motives, plain-language communicators must continuously consider their ethical responsibilities to their audiences:

> If, in our plain language work, we fail to acknowledge the deeply moral and rhetorical nature of the documents we deal with—and this requires thinking about the power, social relations and the long chain of consequences set in motion on publication of a document—if we're not prepared to do that, then we're not what we pretend to be—informed moral citizens involved

in building a better polis—but mere technical assistants in the technopolis, advisors on style and document design, unwitting members of the cult of expediency. (Mackinnon 1997, para. 30)

Mackinnon (1997) closed with seven suggestions for working ethically in plain language. These include questioning the "windowpane" view of language, acknowledging the ethical implications of working with plain language, acknowledging the limits and complexity of plain language, contextualizing one's writing expertise, considering deeply the rhetorical and ethical aspects of plain language, naming a document's author or authors whenever possible, and defining the audience for public documents. This rhetorical view of ethics involves understanding thoroughly the audiences and contexts for plain-language documents.

Another means of understanding the ethics of plain language is in the ways it helps citizens understand and abide by the law. In *A Plain Language Handbook for Legal Writers*, Mowat (1999) devotes a chapter to the ethical foundations of plain language. She notes that the plain-language movement advocates for information in forms that consumers and citizens can readily understand. Because complex, convoluted legalese is unfamiliar to most citizens and difficult for them to understand, it isolates people from their rights and obligations. Effective plain language, however, reflects empathy for readers and provides equitable access to the law (26). Mowat points out that although many readers have low or limited literacy skills, low literacy should not exclude them from rights of justice and citizenship: "Empathy for readers and its related access to the law form the ethical core of plain language" (27). Mowat rejects the idea that legal language is only for specialists, and she faults law firms that draft legal documents without considering their nonexpert audiences. She writes that the best legal writing reflects the ethical assumption that "access to justice means accepting the public's right to understand" (29). Plain-language laws and policies extend citizens' freedoms; plain language bolsters the authority of law and respect for the justice system (151). The public's right to understand the law coincides with the responsibility to follow the law.

A third view of plain-language ethics is that communicators have a responsibility to understand how to reach nonexpert audiences effectively. Writers promoting this view of ethics include Osborne (2005) and Lauchman (2010). In her book on communicating health information, Osborne (2005) discusses what she calls the ethics of simplicity. She writes that in many ways, health professionals are translators for their audiences. They must convey "complicated, rapidly changing, numbers-based information" in a way that is clear and simple enough for nonexpert audiences to understand (51). Osborne's ethics of simplicity involves balancing the writer's responsibility to provide accurate, thorough, and helpful information with the reader's abilities to understand and act on it. Osborne notes that this view of ethics has more to do with choosing the right course of action as a writer than about taking an ethical stand. "Still, it is up to each of us to make a personal commitment to communicate health information in ways that are clear, simple, honest, and complete" (52). Osborne's view of plain-language ethics centers on meeting the audience's needs; communicators do ethical work when they help audiences to understand specific information. Similarly, Lauchman (2010) asserts that workplace writers have a duty to be clear and to help readers understand: "This is a moral and ethical obligation, and every decision (from word choice to format) must spring from it" (18). Lauchman adds that workplace writers must understand "how language engineers meaning" or else their ethics will be irrelevant (18). At the same time, audiences must read "in good faith" and assume a good-faith effort from the writer (19). For both Osborne and Lauchman, the writer's ethical responsibility to be clear and appropriately plain comes from recognizing the audience's needs for information.

Organizations in the US Supporting TPC Practice and Scholarship

Plain-language organizations provide valuable forums for discussions about plain-language work, but their discussions of ethics have not been robust and well documented. Fortunately, organizations supporting TPC scholarship and practice have helped provide ethics literature that is relevant to the practice of plain-language communication. Technical communicators in the US have written and talked about the ethics of their work from the earliest days of their professional societies. In the US, jobs for technical writers initially increased during and following the two World Wars. The military and the manufacturing, electronic, and aerospace industries needed technical manuals and documentation. The Society of Technical Writers and the Association of Technical Writers and Editors each formed in 1953. These organizations merged in 1957 to form the Society of Technical Writers and Editors (STWE). While the STWE was based on the East Coast, the Technical Publishing Society began in 1954 in Los Angeles (Malone 2011). STWE merged with the Technical Publishing Society in 1960 to create the Society of Technical Writers and Publishers. In 1971, this organization became the Society for Technical Communication, now known as STC (Society for Technical Communication 2012a). STC publishes a magazine for technical communicators, *Intercom*, and a research journal, *Technical Communication*. Over time, STC has become an international organization with members and organizational communities around the world.

Although membership in professional associations has declined noticeably over the past decade in the wake of difficult economic times (Sladek 2011), these associations provide valuable professional development opportunities to their members. STC seeks to serve the global community of technical communicators (Society for Technical Communication 2012c) and in early 2013 had more than 6,200 members. The Professional Communication Society (PCS) of the Institute of Electrical and Electronics Engineers (IEEE) has approximately 800 members and promotes communication in science and engineering fields. The IEEE PCS has published its research journal since 1957. First called *Transactions on Engineering Writing and Speech*, the journal is now *IEEE Transactions on Professional Communication*. SIGDOC is the Special Interest Group on the Design of Communication in the Association for Computing Machinery (ACM) and has around 200 members. SIGDOC published the *Journal of Computer Documentation* for several years and recently announced a new journal, *Communication Design Quarterly*. These three associations support the work of technical communicators most directly. While some technical communicators join other organizations such as the American Medical Writers Association, the Association for Business Communication, the American Society for Training and Development, and the User Experience Professionals Association, publications and programs most closely associated with TPC—and TPC ethics—frequently come from STC, IEEE PCS, and SIGDOC.

Foundational Principles of Ethics in TPC

The ethics tradition dates back many centuries. Markel (1997, 2001) identifies foundational theories of ethics as top-down, deductive approaches that apply broadly to many situations. Dombrowski (2000a), in *Ethics in Technical Communication*, and Markel (2001), in *Ethics in Technical Communication: A Critique and Synthesis*, identify a common set of foundational ethical ideas important to TPC, including Kantian rights and obligations, utilitarianism, ethics of care, and—to a lesser extent—virtue ethics.

Kantian Rights and Obligations

As noted in chapter 1, the plain-language movement has focused considerable attention on consumers' and citizens' rights to receive information they can readily understand. Immanuel Kant, an eighteenth-century German philosopher, took a reason-focused approach to ethics, and his approach has influenced discussions of ethics for more than two centuries. In *The Foundations of the Metaphysics of Morals*, Kant ([1785] 1969) describes an approach to ethical behavior based on reason. Kant's central principle is that nothing is good unless it comes from goodwill—the will to act from duty. Acts motivated by self-interest have no moral worth, but acts motivated by duty do have it (Markel 2001, 48). As Dombrowski (2000a) notes, a system like Kant's emphasizing obligation or duty is a deontology (47). As the following discussion of Kant's perspective shows, Kant links ethical reasoning with action.

In the *Foundations* ([1785] 1969), Kant describes two approaches to action. The hypothetical imperative is contingent on circumstances and desires: as Markel (2001) describes it, "If you want to accomplish x, then do y" (48). On the other hand, the categorical imperative describes what a person should do regardless of contingencies: "Do y." The categorical imperative reflects careful reasoning and is absolute: "It does not admit qualifications, explanations, or exceptions" (Markel 2001, 48). Dombrowski explains how the strict categorical imperative aligns with a person's free will: "We usually think of duty as an onerous burden and a limitation of our freedom. But, paradoxically, Kant's theory of duty is founded on the radically autonomous free will and its capacity to choose otherwise and on its reasoned self-persuasion not to choose otherwise. It is duty based in freedom" (Dombrowski 2000a, 49). Similarly, communicators may choose whether to communicate with their audiences in plain language; while some laws mandate plain language in certain cases, no universal imperative for it exists. If communicators choose to use plain language because they believe the rhetorical situation warrants it, in a way they do so out of duty.

Kant's *Foundations* provides three versions or formulations of the categorical imperative. The first is this: "Act as though the maxim of your action were by your will to become a universal law of nature" (Kant [1785] 1969, 421). According to this formulation, our actions should be appropriate and consistent in all settings, universally. For example, telling the truth in all circumstances would be following a categorical imperative that makes sense as a universal law of nature.

The second formulation emphasizes each individual's rights: "Act so that you treat humanity, whether in your own person or in that of another, always as an end and never as a means only" (Kant [1785] 1969, 429). For many, this second formulation resembles the Golden Rule found in Christian teaching and other religious traditions: Do to others as you would have them do to you (Matthew 7:12). Markel (2001), however, shows that Kant did not simply repeat the Golden Rule; he points out that Kant's second formulation does not mention any reciprocity (51). Kant says in a footnote that this maxim cannot become a universal law "because it contains the ground neither of duties to one's self nor of the benevolent duties to others" (Kant [1785] 1969, 430n). Kant appears to see in the Golden Rule the possibility of quid pro quo behavior, acting a certain way in order to get something. This view of the Golden Rule conforms to a hypothetical imperative but not to the categorical imperative, which Markel calls the bedrock of Kant's ethics. Markel (2001) comments that if we are to treat others as ends and not means, we must accord them their full dignity and write and speak truthfully (51–52).

Kant's third formulation of the categorical imperative, like the first formulation, focuses on individuals identifying universally appropriate behavior. He writes that it is "the idea of the will of every rational being as a will giving universal law. A will which stands under laws can be bound to this law by an interest. But if we think of a will giving universal laws, we find that a supreme legislating will cannot possibly depend on any interest" (Kant [1785] 1969, 432). An interest, in this case, would be the self-interest of acting a certain way with the expectation of reciprocity. Markel (2001) explains that the third formulation summarizes the first two and describes a universal realm where people live by "self-created rules derived from reason" (54).

The technical communication literature provides two complementary ways of understanding Kant's categorical imperative. While Dombrowski (2000a) describes Kant's categorical imperative in terms of obligations, Markel (2001) describes it in terms of rights. In Kant's approach, the rights of one person define the obligations of another. Respect for individual rights resonates with the plain-language movement because many practitioners believe people have a right to clear information. Chapter 3 presents practitioners' views on this right, which many practitioners believe is inherent; however, some practitioners question whether audiences do have this right at all. Kant's second formulation, to treat humanity as an end but never as a means to an end, includes a respect for people that carries through Buber's dialogic ethics, discussed later in this chapter.

Markel (2001) points out that some aspects of Kant's theories do not stand up to intense scrutiny. For example, Kant assumes free will but does not prove that free will exists. Some ideas are subject to counterexamples and paradoxes. In another example, Markel refers to Kant's essay "On the Supposed Right to Lie from Altruistic Motives," wherein Kant says that if an inquiring murderer is looking for your neighbor and you know where the neighbor is, you do not lie to the murderer because of the categorical imperative to tell the truth in all circumstances. Needless to say, Markel (2001) does not agree with Kant's reasoning in this case with an individual's life at risk (45). Furthermore, Kant does not account for special relationships such as friendships and family and the ways in which we might act differently because of our relationships. For example, a parent might choose not to turn in a thief to the authorities if the thief is his or her own child, if that consequence might harm the child more than helping. Markel (2001) writes, "All this being said, [Kant's] three main ideas—that principles have to be supported by reasons, that they must apply consistently to everyone, and that people must be treated as ends and not merely as means—remain powerful as general ethical guidelines" (56). Whether individuals agree with all of Kant's ideas or not, they have value as foundational principles to consider when seeking an ethical course of action.

Utility

Utility refers to the positive effects of an action, and the utilitarian perspective, promulgated by Bentham and Mill, promotes the greatest good for the greatest number of people. Dombrowski (2000a) calls the utilitarian perspective a technological approach to ethics (54); Markel (2001) calls it "the ethics of pragmatism and common sense" (58). Societies that depend on a myriad of technological systems respect utility and a parallel value, efficiency. The utilitarian value of plain language is clear. Advocates have for decades noted the time and money saved when people receive plain-language documents they can use efficiently (e.g., Kimble 2012). However, if the value of plain language becomes merely an amount of money listed on a balance sheet, it is no longer a humanistic concern. Utility is one aspect of the ethical value of plain language, but not the only one.

Bentham, writing in late eighteenth- and nineteenth-century England, saw utility as a measure of pleasure for all those affected by an action, not just the agent. He promoted a pleasure-focused or hedonistic calculus to determine the utility of each action. This calculus includes judgments about intensity, duration, certainty as opposed to uncertainty, propinquity (closeness) as opposed to remoteness, fecundity (the chance of recurrence), purity (the chance that the opposite won't follow it), and extent or the number of people affected. Many are likely to agree with Markel's (2001) assertion that hedonism is a flawed theory of value: we value many things that do not reduce to pleasure. Bentham's hedonistic calculus is unworkable because most effects are not measurable and we cannot see into the future (60–62). Similarly, most communication situations are difficult to view through a hedonistic lens. In BUROC situations in which individuals face important decisions amid stress and uncertainty, plain language can be helpful and effective, but it cannot guarantee feelings of pleasure or happiness.

John Stuart Mill viewed utility as happiness. Happiness brings pleasure and absence of pain; unhappiness brings pain and removes pleasure. In *Utilitarianism*, published in 1863, Mill writes that higher pleasures are intellectual pleasures, while lower pleasures are closer to our human appetites. Sanctions or corrections of our behaviors, from our consciences as well as our environment, encourage us toward good conduct. Markel identifies some problems with Mill's theory and with utility in general. He says that Mill's discussion of "The greatest good for the greatest number" is ambiguous, the discussion of qualitative aspects of pleasure is imprecise, Mill's discussion of motives for altruism is vague, and conscience is not an effective regulator of behavior for all people (Markel 2001, 63–65).

Utilitarianism has developed into the underlying force behind cost–benefit analyses so common in contemporary organizational life (Markel 2001, 57–58). While utility often provides a useful and helpful perspective, Dombrowski (2000a) describes two utility-based decisions about airplane safety that had tragic results. His first example is the ValuJet plane that crashed in the Florida Everglades in May 1996. A fire in the cargo hold—caused by ignition of an improperly shipped oxygen generator—went undetected until it was too late to respond. The Federal Aviation Administration (FAA) had previously decided that the relatively few benefits of fire-safety technology did not offset their high costs (55). After the crash, the FAA mandated installation of fire detection and suppression systems by March 2001 (Sarkos 2011, 11). Dombrowski's second example is TWA Flight 800, which exploded over Long Island Sound in July 1996. The probable cause was fuel vapors ignited by a short circuit; all 230 passengers of Flight 800 perished. Dombrowski (2000a) writes that the FAA received criticism for focusing on costs over benefits when deciding not to require fuel-vapor control or purging systems (55). In 2008, the FAA required engineering changes to increase the safety of airplane fuel systems. These changes apply to central fuel tanks and not wing tanks. The FAA had proposed design changes in 2005, but the aviation industry balked over costs (Lowy 2008). While plain language often benefits organizations and their constituents (e.g., Kimble 2012), organizations must invest time and money to produce plain-language documents. If the benefits of plain language exist only in terms of finances, a plain-language effort is no different from any other cost that an organization can cut from its budget.

While utilitarianism has limitations as an ethical theory, its pragmatic perspective is useful. Writes Dombrowski (2000a), "Nevertheless, sometimes the utilitarian approach, despite its impersonalness and other difficulties, seems to be the only reasonable ethical approach to take. Medical ethics provides a good example of the validity of this approach in some situations" (55–56). Dombrowski gives the example of trying to allocate three organs for transplant among ten eligible patients; each patient would benefit from a transplant, but some may benefit more than others. Markel (2001) argues for considering utility alongside other core values: despite the "considerable attractiveness of utility, it cannot exist as a coherent ethical principle without some consideration of rights and justice" (72). Taken to an extreme, utilitarianism could reduce people, their feelings, and their experiences to mere amounts of money in a budget; decisions on actions could become exercises in mere addition and subtraction. Utility is an important view of ethics, but unchecked it could lead to undesirable consequences.

Care and Feminist Approaches

While plain language can provide measureable benefits that support utilitarian goals, it also provides other benefits that are harder to measure but no less valuable. For example, the ethical concept of care, stemming from feminist approaches to psychology and philosophy, provides a perspective on ethical behavior that focuses on interpersonal relationships—which perspectives such as utility and Kantian deontology do not address. In recent decades, scholars have acknowledged that men and women often perceive moral questions differently. Women tend to emphasize creating and maintaining relationships, to focus on specific details in ethical situations, and to deemphasize abstract principles; men tend to view the details from a more impersonal, distant stance and to more greatly value abstract principles that apply broadly (Dombrowski 2000a, 62; Markel 2001, 89). The respect for interpersonal relationships in feminist approaches complements Buber's emphasis on genuine, reciprocal dialogue, which appears later in this chapter.

Feminist approaches provide alternatives to male-influenced views of ethics that do not account for differences in men and women. In *In a Different Voice: Psychological Theory and Women's Development*, Gilligan (1982) critiques Kohlberg's work on moral development. Kohlberg, focusing on abstract ideas such as rights and justice from a Kantian perspective, followed the development of 84 boys over 20 years and identified six stages of moral development. Through a series of studies, Kohlberg judged that most men reached the fourth stage of development while most women reached the third. Kohlberg's work led some to consider women less ethical than men. Gilligan (1982), on the other hand, notes that girls take different paths than boys as they mature. Over time, men tend to focus ethical judgments on individual rights and on abstract ideals like justice while women view ethical questions through their relationships with and responsibilities toward others—that is, through the perspective of caring. Gilligan (1982) shows that women are not deficient to men in their moral development, but that women's development differs from men's. From a feminist perspective, communicators might choose plain language for audiences facing BUROC situations because it helps them understand and act upon information to cope with a stressful, difficult situation and not just because plain language might lead to better results from a utilitarian perspective.

Feminist approaches respect individuals' needs for interaction and dialogue. In *Caring: A Feminine Approach to Ethics and Moral Education*, Noddings (2003) describes caring as a feminine view of ethics that emphasizes receptivity, relatedness, and responsiveness. The one doing the caring and the one who is cared for share a relation with each other; ethical judgments of care are specific to relationships and need not be universalizable. The ethic of caring is not tender-minded, but is tough; it embodies tenacious advocacy of ideals and sacrifice for the one who is cared for (3–5). While some question whether we should consider care a foundational ethical approach, we should think about the ways people experience and apply the ethic of care (Bowden 1997). Markel (2001) writes that care has great potential to "redress the imbalance of foundational ethical approaches, which place too little value on personal and familial relationships" (109). A greater respect for interpersonal relationships could lead people toward better, more genuine dialogue.

While the concept of care provides one feminist approach to ethics, it is not the only approach. After reviewing the literature, Dombrowski (2000a) writes that a feminist perspective on ethics in technical communication brings attention to important factors that escape notice through conventional ethical approaches. It reveals the bias against women and women's work in a variety of ways in technical communication; it questions the importance of objectivity as a preeminent value while asking whether objectivity is even attainable (62). Many agree with Walker (2001) when she writes that feminist ethics examine the morality of specific distributions and exercises of power (4). In chapter 3, several practitioners make points about the social value of plain language that reflect feminist points of view. In chapter 4, plain-language professionals at the company Healthwise note that plain language can address the imbalance of power between physicians, who are experts, and their novice patients.

Other Approaches to Ethics

Because the ethics tradition goes back many centuries, discussions about ethics must necessarily focus on some approaches and exclude others. While Markel

(2001) and Dombrowski (2000a) agree on some key concepts, each also identifies some other important ethicists and approaches. Dombrowski (2000a) discusses Confucian ethics, Emmanuel Levinas's outward-looking ethics focused on the Other (to which I will return later), and Bernard Gert's approach to ethics and morality. Markel (2001) discusses G. E. Moore and his views on what is good, W. D. Ross and his discussion of prima facie duties, John Dewey's experimental ethics, James Rachels's list of higher-order considerations for determining ethical action, John Rawls's principles for a fair society, and Jürgen Habermas's discourse ethics.

Virtue Ethics

Virtue ethics, like care, focuses on the character of the agent. It emphasizes aspects of character that let a person make good decisions (Markel 2001, 98). Swanton (2003) explains that in virtue ethics, "conceptions of rightness, conceptions of the good life, conceptions of 'the moral point of view' and the appropriate demandingness of morality, cannot be understood without a conception of relevant virtues" (qtd. in Baron, 2011, 28). Baron (2011) and others advocate understanding virtue ethics as a genus or category containing different species of approaches to virtue ethics. The most prominent species of virtue ethics is most likely that of Aristotle and his treatise the *Nicomachean Ethics* (1975). As Dombrowski (2000a) describes Aristotle's position, ethics is the study of what is involved in good actions. Ethics is about seeking goodness for its own sake; the good is intrinsically good and right (41).

In Book II, Chapter VII of *Nicomachean Ethics*, Aristotle lists 12 individual virtues of character: courage, or bravery; temperance; liberality with large sums, or generosity; magnificence with small sums; greatness of soul, or proper pride; appropriate ambition; good temper, or mildness; truthfulness; wittiness; friendliness; modesty; and righteous indignation (1975, 97–105). Markel (2001) summarizes Aristotle's main points:

- Ethics is a branch of politics and is meant to be practical.
- Ethics is imprecise and nonprincipled (i.e., it does not provide principles to determine an ethical course of action).
- Virtue is the means of achieving happiness.
- Acting virtuously is self-reinforcing.
- Acting virtuously consists of finding the golden mean between extremes (e.g., courage is the midpoint between foolhardiness and cowardice).
- The highest form of activity is the contemplative life. (99–100)

Although ethicists such as Alasdair MacIntrye argue for virtue ethics, Markel (2001) finds the approach limited. He identifies three major criticisms of virtue ethics: it does not accommodate change in a community, it focuses on individuals, and it offers no clear guidelines for decision making. Markel cites Kohlberg, who derides the approach as a "bag" of virtues lacking order and principles of application (103). Virtue ethics focus on an individual's thoughts and choices; plain-language communicators need to focus their attention outward, on their audiences. Virtue ethics are an important part of the ethical tradition, but they have limited relevance to plain language.

Professional and Academic Perspectives on Ethics in TPC

We now have several decades of writing about ethics in technical communication. Over this time, authors and editors have collected articles and papers through review essays, anthologies, and special issues or sections of journals. In 1980, STC published a special section on ethics in *Technical Communication*. *IEEE Transactions on Professional Communication* published special issues on ethics in 1987 and 1995. Brockmann and Rook (1989) edited STC's anthology, *Technical Communication and Ethics*. Doheny-Farina (1989) reviewed three decades of technical communication ethics literature in an edited collection on technical and business communication. Sanders (1997) followed with a review of articles for the collection *Foundations for Teaching Technical Communication*. One review essay by Dombrowski (2000b) also addresses technical communication ethics in the last part of the twentieth century; a second essay (Dombrowski 2007) focuses primarily on articles published after 2000. Markel (2001) surveys the literature in his book on technical communication ethics. *Technical Communication Quarterly* published a special issue on ethics in 2001; a 2004 special issue of *IEEE Transactions on Professional Communication* on case studies of technical communication highlights many points for discussion about ethics. Other articles on ethics have followed. STC's *Intercom* magazine periodically publishes ethics cases and occasionally publishes readers' responses to them.

In general, articles on ethics in technical communication tend to represent two of three perspectives described by Clark (1987): the professional perspective focused on immediate and practical workplace contexts, and the academic perspective concerned with broader principles and theoretical approaches to ethics. I will show later in this chapter that Clark's third perspective promotes a focus on dialogue within communities. Dombrowski (2000b) uses Clark's two perspectives in his first review essay. Sanders (1997) groups academic articles under two headings: philosophical, relating to ethical approaches from studies of philosophy, and rhetorical, relating to ethics of composing messages for specific audiences. My goal below is not to repeat any of the existing work on ethics, but to identify important perspectives on ethics in technical communication. In some cases it is difficult to say that an article is wholly academic or professional: the categories sometimes overlap, and many writers of professional articles are academics. Some articles could fit in either category.

The Professional Perspective

The special issue of *Technical Communication* on ethics published in 1980 discusses three approaches to ethics: professional, legal, and moral. Brockmann and Rook (1989) republished key articles from that issue in their anthology. Radez (1989) encourages STC members to focus on professional ethics, not the legal or the moral. Radez's ethic of professionalism includes doing quality work, cooperating effectively, giving credit and recognition when it is due, avoiding impropriety with publishers and advertisers, and avoiding plagiarism (4–5). Shimberg writes that all three approaches to ethics are valuable, although a writer's professional ethics will be most evident and are most important. Legal ethics are most clear, but we tend to test them least often; moral ethics are labyrinthine and complex, and senior technical writers are much more likely to apply them on the job than are junior writers (1989b, 11–13). Sachs writes that it is difficult to separate ethics into professional, legal, and moral categories, arguing that moral ethics underpins the other two (1989, 7). Sachs asserts (1989, 9) that moral ethics is based on the Golden Rule (doing unto others as you would have them do to you). Sachs offers nine cases for consideration but does not say which perspective is most appropriate for each case. In some cases, communicators must follow laws that mandate certain levels of clarity and readability in content. Most practitioners who contributed their views of ethics to chapter 3, however, tend to have views that fit in the category of professional ethics; some professionals shared views in the moral or philosophical category.

In addition to the lenses on ethics provided by the professional, legal, and moral perspectives, some professional sources briefly allude to philosophical views on ethics. Writing for a professional audience, Wicclair and Farkas (1989) briefly describe goal-based (or utilitarian), duty-based, and rights-based ethics. In a chapter of his book on technical writing style, Jones (1998) describes ethics based on obligations, ideals, and consequences. Citing ethicist Vincent Ruggiero, Jones says the principle of respect for persons underlies almost all ethical systems (1998, 241).

Other threads of conversation within professional publications on technical communication include the need to communicate clearly and honestly, practical ethics on the job, ethics and law, codes of ethics for technical communicators, ethics and technical marketing, and ethics and visuals.

Communicating Clearly and Honestly

Professional ethics involves writing clearly and truthfully. Shimberg (1989a) identifies four ethical traps technical communicators are likely to face: imprecision and ambiguity, or technical doublespeak; understating or suppressing the truth; overstating the negative; and semanticism, or word choice that readers might misunderstand. Griffin (1989) discussed the relationship between rhetorical choices and ethical choices in the role to which a writer commits. According to Griffin, writers taking on roles of experts should make rhetorical choices consistent with their ethical obligations. Arthur E. Walzer (1989a) states that when "existing practice" is the standard, some might think that whatever is rhetorically effective is right and ethical. He gives the example of Robert C. McFarlane, who testified twice before Congress about the Iran Contra affair in the 1980s. MacFarlane intentionally gave statements that were misleading but not untrue; he fostered false inference among his hearers (150–51). Walzer cites Grice's maxims of conversation to show how writers can avoid unethical effects of false implicature. Riley (1993) expands on Walzer's article, providing more background on the sociolinguistic principles behind implicature and analyzing cases for discussion—including two originally from Walzer (1989a). Writing about the thread of clear-and-honest communication in technical communication literature, Markel (2001) notes that the philosophical backing behind such articles is largely implicit and receives little explicit attention (12–13). Nevertheless, one common theme among these articles is a focus on the audience's experience in communication. Ethical behavior leads audiences to truth and to accurate understandings; unethical behavior leads to inaccuracy and misunderstandings.

Practical Matters of Ethics

Michaelson (1990) identifies violations of ethical behavior in the technical communicator's work life that are sometimes intentional, sometimes unintentional. These include sins of omission, such as hiding defects, withholding contradictory information, and failing to credit the work of others; unfair bias, including failure to cite sources that challenge one's supporting sources, cropping photographs to mislead the reader, and overstating positive information; ambiguity, including thinly veiled speculation, vague writing that invites misinterpretation, and incomplete citations; plagiarism of others' work and one's own; and indiscriminate publication of pieces that fail to meaningfully advance knowledge in a field. Markel (1991) also identifies common ethical challenges technical communicators face; these include avoiding plagiarism, maintaining trade secrets, and deciding whether to take action as a whistleblower in response to egregious ethical violations. While some of these practical matters do affect the audience, such as cropping a photograph to create a false impression, most result from choices an individual makes that reflect poorly on him or her. These practical matters focus on the individual's relationship to others within the organization and within the professional communication community.

Ethics and the Law

The professional view that ethical behavior involves following relevant laws is evident in an article by Shimberg (1989b) and in the special issue of *IEEE Transactions on Professional Communication* edited by Doheny-Farina (1987).

Doheny-Farina's (1989) book chapter also identifies many articles in this vein. Herrington (2003), who holds doctorates in both jurisprudence and technical communication, writes that the law and ethics have different purposes: the law exists to dispense with or avoid conflict while ethics exist to guide personal conduct and behavior. While following the law tends to lead to ethical behavior, it is not a guarantee. Some laws, such as those that supported slavery and turned slaves into property, clearly are not ethical. Herrington (2003) recommends applying the axis-of-power test to determine how to communicate ethically. According to this test, those who have power to communicate information to others and do so (by choice or by obligation) also have a responsibility to communicate honestly, without masking information or misleading people. If withholding certain information will mislead readers or conceal information vital to them, then the communicator cannot pass the axis-of-power test in that scenario (11). The axis-of-power test brings to mind Walker's (2001) comment that feminist ethical perspectives examine specific distributions and exercises of power. The BUROC model for identifying opportunities to use plain-language covers many situations in which individuals lack stature, power, or agency in their dealings with bureaucratic organizations. Plain-language communicators can help to ameliorate the imbalance of power that individuals experience.

Codes of Ethics for Technical Communicators

A code of ethics can encourage a group's members to behave ethically, but it cannot take the place of each member's internal ethical compass. The history behind the STC's code of ethics, a document now titled "Ethical Principles," appears in works by Malone (2011) and Walzer (1989b), among others. Malone identifies a 10-item ethical code created by Robert T. Hamlett, the first president of the Association of Technical Writers and Editors. It contains 10 affirmative statements about good professional behavior for technical writers (Malone 2011, 293). It did not appear to have wide influence. After the merger of TWE and the Society for Technical Writers created the Society for Technical Writers and Editors, the STWE sought to increase its standing as a profession. In 1958, the STWE approved its Canons of Ethics, which were adapted from those of the Engineering Council for Professional Development. Malone (2011) and Brockmann (1989a) each identify opinions calling for true professions to value ethical behavior. Because the Canons contain general statements about professional behavior and do not mention writing, one might get the impression that it was more important to STWE simply to lay claim to a code of ethics rather than to guide the behavior of its members (Walzer 1989b, 101).

In a piece written before the Watergate scandal, Perica (1972) describes the American public's weariness of lies and partial truths from politicians, government officials, and the media. Noting pervasive pressures to put company profits ahead of ethical behaviors, he calls for a code of ethics to guide the relationships between technical writers and their customers (6). Malone (2011) writes that the Watergate scandal provided motivation in the late 1970s for STC members to revisit their code of ethics. Articles by Shimberg (1989b), Radez (1989), and Sachs (1989) document some of the thinking of the 1970s and 1980s. Brockmann and Rook (1989) allow readers to compare the STWE's Canons of Ethics adopted in 1958 and the STC's Code for Communicators revised in 1988, along with then-current codes from the International Association of Business Communicators, the American Medical Writers Association, and the International Council for Technical Communication (INTECOM). Although the IEEE PCS does not have a code of ethics for its members, the IEEE expects all members to follow the IEEE Code of Ethics. Neither Plain Language Association International nor Clarity International, the two largest associations supporting plain-language professionals, has a code of ethics for members.

STC last updated its current code, titled "Ethical Principles," in 1998. It calls on technical communicators to uphold principles of legality, honesty, confidentiality, quality, fairness, and professionalism. The section titled "Honesty" has the most relevance to plain-language communicators. The first sentence calls for technical communicators to show concern for others and be aware of their work's consequences: "We seek to promote the public good in our activities." The second sentence calls for honesty, a bedrock of the field's professional ethics: "To the best of our ability, we provide truthful and accurate communication." The third sentence points toward the ways clear communication benefits audiences and the importance of meeting audience needs and expectations: "We also dedicate ourselves to conciseness, clarity, coherence, and creativity, striving to meet the needs of those who use our products and services." The fourth sentence implies that unclear communication causes problems for audiences: "We alert our clients and employers when we believe that material is ambiguous." (Society for Technical Communication 2012b). The remaining four sentences in the section address citing others' work properly, giving bylines to only those who contribute to a document substantively, using facilities and resources only with permission, and using only truthful advertising. In short, this section of the STC's code encapsulates much of the field's professional perspectives on communicating clearly and honestly and on the practical matters of ethics as a colleague and employee.

After reviewing several codes of conduct, Buchholz (1989) writes that ethical codes for communicators should express clear attitudes about communicators' relationships to the public and about the nature of truth. Practitioners need help in deciding which public most closely represents the code's ideal public, and they need guidance that is detailed enough and flexible enough to be helpful when forces seek to obscure the truth in large or small ways. Buchholz provides six recommendations for creating codes of ethics that are expository, analytical, and evaluational. Among these are providing definitions, offering examples and illustrations, analyzing especially difficult cases, and identifying issues subject to interpretation (65–68). Markel (1991) notes that self-interest often compels employees to maintain silence about company problems. Because codes of conduct must be flexible to cover many situations and thus may be vague, Markel says some see codes of conduct as public-relations tools and not as true guides of conduct. None of the plain-language professionals in chapter 3 mentioned a professional code of conduct as an influence on his or her ethics.

Ethics and Technical Marketing

Many professional technical communicators write marketing documents, and some articles address the specific ethical challenges that marketing communicators face. Bryan's (1992) article, "Down the Slippery Slope: Ethics and the Technical Writer as Marketer," identifies some of the ethical challenges associated with writing for marketing in competitive environments. Bryan suggests that marketing writers ask themselves what they would write if readers could hold them—not their employers or supervisors—personally accountable for their work (87). Porter discusses the ethics of technical advertising. He discusses the case of an insurance company that went to court for misrepresenting certain policies as enduring for the policyholder's lifetime. Porter (1987) identifies key ethical principles for those who write as advocates or marketers, and he discusses the relationship between ethics and readability. Porter states that writers have an ethical responsibility to know the truth about a product or service and to communicate that truth to consumers (188).

In their textbook on technical marketing communication, Harner and Zimmerman (2002) distill their advice in two admonitions: do not lie, and do not steal (64). The authors advise writers to tell the truth and to avoid promoting nonexistent features. They say stealing someone else's trademarks, taglines, and intellectual property is theft; they urge writers to respect confidentiality agreements and to avoid collusion on prices. Harner and Zimmerman tell readers to "carefully establish a personal standard of ethics and integrity long before you're challenged to test it" (65) but do not provide further guidance. Many plain-language communicators serve a variety of clients, and some create marketing documents. No matter who their clients or audiences are, plain-language communicators are likely to encounter ethical dilemmas on the job.

Ethics and Visuals

Because visuals (tables, graphs, figures, photographs) can mislead readers and create false impressions, scholars have argued that communicators should produce visuals ethically, without any intent to deceive. Like technical communicators in a variety of industries, plain-language communicators use visuals to communicate with their audiences. Bryan (1995) identifies seven types of distortion in charts and graphs that can deceive readers. Allen (1996) discusses ways in which changes to photographs and graphic icons can mislead readers; she argues for visual rhetorics that can inform the ethical choices designers must make. Herrington (1995) reviews tables in a report on the standoff near Waco, Texas, between the Bureau of Alcohol, Tobacco, and Firearms (ATF) and the Branch Davidians. Herrington (1995) shows that seemingly small changes in type size and spacing had the cumulative effect of emphasizing the casualties among ATF personnel while making the Branch Davidians' casualties appear less significant.

One survey of 500 technical communicators and 500 technical communication teachers asked whether each of seven hypothetical changes to a visual in a document was an ethical change (Dragga 1996). Respondents said changes leading to greater deception and greater injury to the reader were unethical, and they said changes not leading to deception or injury were generally ethical. Comments in the survey responses often focused on the consequences of each action; this reflects a goal-based view of ethics—in this case emphasizing what to avoid over what to do (Dragga 1996, 262–63).

Another area of discussion about ethical visuals involves depicting data about human injuries and deaths. Dragga and Voss (2001) argue that it is unethical to present statistical, visual, or photographic evidence (2003) about deaths and injuries to humans in any way that hides the victims' humanity. The authors show ways to use line drawings, captions, victims' names and biographical information, and other graphic devices to underscore the fact that the situations depicted affected real people. Lancaster (2006) writes about fatalgrams, which depict accidental deaths using photographs or line art. While government agencies use fatalgrams to understand and prevent industrial accidents, Lancaster cites examples of private citizens using these images (which are US government works without copyright restrictions) in insensitive, crass attempts at humor. Lancaster describes (2006) how an ethic of care could support proper use and avoid misuse of fatalgrams.

Data displays—from tables and bar graphs to interactive online maps and diagrams—exist to help readers understand the data behind them, and thus clarity is a virtue. Kostelnick (2008) says the "rational, efficient rhetoric of data design" developed over several centuries "embodies an intrinsic ethical component because it implies that readers deserve a full, unadulterated disclosure of the data and that designers have a moral imperative to provide it" (118). Ultimately, Kostelnick writes, the clarity of a data display will be determined not by how well it meets abstract principles or heuristics but by how well readers understand the data when they view the display. As new approaches to data display are developed, the socially contingent nature of clarity will likely become more apparent to technical communicators and plain-language communicators alike (128).

Amare and Manning (2013) use philosopher C. S. Peirce's concepts of decoratives, indicatives, and informatives to develop a theory of information design uniting visuals, texts, and ethics. Decoratives evoke feelings, indicatives provoke action, and informatives promote understanding. Amare and Manning use a triangle diagram to show the relationships between Peirce's concepts. The authors place decoratives in one corner, indicatives in the second, and informatives in the third. Each side of the triangle is a continuum between the two concepts on each end. Along each side of the triangle, Amare and Manning (2013) place four groups of communicative tools or practices; a tenth group of visual communication tools appears in the center of the triangle (9). The authors then explain how Peirce's three fundamental categories of perception and understanding provide a grammar for understanding information design (24). Firstness, raw sensation and perception, corresponds with decoratives. Secondness, action and reaction, corresponds with indicatives. Thirdness, patterns based on principle or habit, corresponds with informatives. Secondness contains firstness, and thirdness contains both firstness and secondness. Amare and Manning write that in Peirce's system, the communicator and the audience define what is ethical by agreeing on shared goals and purposes. At the same time, any agenda that is self-defeating or unsustainable is unethical, even if all parties agree to it. An agreement to suppress evidence of drug side effects cannot be ethical, for example. Strictly utilitarian views that do not consider long-term sustainability also fail in Peirce's ethics. "This, in summary, is Peirce's epistemological ethics: in one way or another, all ethical goals require the open pursuit of truth and discourage the deliberate hiding of information that a community is deliberately seeking. Visual communication habits of a society may be seen as key indicators of the larger ethical commitments of that society" (Amare and Manning 2013, 14). Plain-language communicators can benefit from Amare and Manning's focus on dialogue with the audience and striving toward shared goals.

Applied Professional Ethics from Allen and Voss (1997)

Much of the professional perspective on ethics shown in the articles above also appears in *Ethics in Technical Communication: Shades of Gray*, by Allen and Voss (1997). Allen and Voss identify 10 values providing a foundation for technical communicators' ethical behavior: honesty, loyalty, privacy, quality, teamwork, avoiding conflicts of interest, cultural sensitivity, social responsibility, professional growth, and advancing the profession (38). While each of these 10 is a value or a virtue toward which technical communicators should strive, the authors describe each as a duty: honesty is our "duty to tell the truth," legality is our "duty to obey the law," privacy is "our duty to respect the rights of others" (38), and so on. Allen and Voss combine values from the world of professional work with a duty- or obligation-focused approach to ethics similar to Kant's.

Among the 10 values, honesty, social responsibility, and cultural sensitivity are most salient to plain-language communicators. To reflect honesty, communicators must not only pursue it but must also avoid the many means of lying. To reflect social responsibility, communicators must create documents that serve and protect users; they must communicate accurately and avoid the obfuscation of doublespeak. To reflect cultural sensitivity, communicators must avoid stereotypes, avoid sexist language, and avoid tokenism while also striving to respect multiculturalism. Allen and Voss (1998) also published an article on ethics for editors based on their book's approach. They write that editors act ethically by valuing honesty in their relationships with authors and by editing to give readers a clear and honest message.

Voss and Flammia (2007) discuss the impact of five of Allen and Voss's (1997) 10 values on intercultural communication. Voss and Flammia discuss cultural sensitivity, legality, privacy, teamwork, and social responsibility in terms of three foundational approaches to ethics—Aristotelian virtue ethics, utilitarianism, and Kantian responsibilities. Voss and Flammia write that cultural sensitivity is the key to ethical intercultural communication. Technical communicators and managers should have more than a superficial knowledge and appreciation of foreign cultures; they should strive to view issues from each audience's cultural perspectives (85–86). The emphasis on respect for other cultures coincides with Buber's (1970) advocacy for genuine dialogue between parties, which I discuss later in this chapter.

The Academic Perspective

While the professional perspective on technical communication ethics has developed over several decades, the academic perspective on ethics has kept pace. Main areas of this academic inquiry include rhetorical approaches, approaches to integrating ethics into curricula, critical theory, feminist approaches, and dialogic ethics. For plain-language communicators, this perspective effectively complements the foundational and professional perspectives—especially in the area of feminist ethical approaches.

Rhetorical Approaches

Many academic articles investigate the relationship between technical communication and the larger field of rhetoric (Markel 2001). This connection to rhetoric extends rhetorical concepts of ethics developed over many centuries toward modern-day applications. In her landmark essay, "A Humanistic Rationale for Technical Writing," Miller (1979) critiques positivist views of science and technical writing. Pointing to the socially constructed nature of scientific knowledge, Miller argues for a "flagrantly rhetorical" approach to technical and scientific writing. Such writing is not merely "a set of techniques for accommodating slippery words to intractable things, but [is] an understanding of how to belong to a community" (617). Sanders (1988) adapts the Rogerian approach to rhetorical argumentation, influenced by the work of psychotherapist Carl Rogers, into a three-step approach for technical writers that emphasizes identifying with the audience's point of view. Sanders seeks a rhetorical approach that emphasizes the ethical challenges technical writers face when they become de facto "anonymous spokespersons for a variety of businesses, research laboratories, and the government" (75). Ornatowski (1992) writes that technical writers are rhetoricians who continually make ethical choices and negotiate between conflicting demands. He cites his interview with an engineer who struggled to describe the problems a piece of equipment had passing a certain test. "The recognition of the fundamental rhetoricity of technical writing," Ornatowski says, is the first step toward integrating ethics into technical writing curricula (92). See Ornatowski (1997) for a more philosophical description of technical communication as rhetorical. Like other rhetoricians, plain-language communicators continually encounter rhetorical situations in which they must craft messages for specific audiences in response to particular exigencies.

Aristotelian rhetoric provides a means of understanding ethics in the context of corporate communication. Kallendorf and Kallendorf (1989) note that some object to Aristotle's *Rhetoric* because he writes that rhetoric allows someone to argue both sides of an issue and to prove opposite statements. They point out, however, that Aristotle also tells readers not to advocate evil. While Plato advocates a view of truth as absolute, Aristotle sees truth as contingent and socially constructed. Thus Aristotelian rhetoric strives to discover and articulate what is *probably* true and just; rhetoric generates "the only truth available in a world of contingency" (57). Kallendorf and Kallendorf write that an ethical rhetorical process is open and invites scrutiny as ideas are articulated. "Under the Aristotelian paradigm, the discussion through which a defensible position is articulated provides a check on what is said and done" (59). As scrutiny of a message increases, any flaws in the message's content or delivery are more likely to be exposed and corrected. Kallendorf and Kallendorf see discussion and debate about corporate messages as key requirements for ethical workplace communication. For many plain-language communicators, especially those writing for audiences facing BUROC situations, the idea of crafting a document with contingent and not absolute truth is likely unsettling. Working collaboratively within their communities of practice and with the audiences they serve, however, plain-language communicators can do thorough, ethical rhetorical work that meets audiences' needs.

Porter (1993a) draws from three of Aristotle's works—the *Rhetoric*, the *Nicomachean Ethics*, and the *Politics*—to develop a postmodern, situational ethics of rhetoric and composition. Porter's model provides a pluralistic ethics that allows rhetors to explore the competing values present in any writing context. He hopes his approach will help writers consider the political and ethical consequences of each act of composition, of what is good and desirable for each specific situation. "[E]very composing event is itself an ethical decision, not simply a presentation of an already preformed ethical position" (223). Plain-language communicators, like technical communicators in general, need to consider the ethical impacts of their work continually and not only in response to isolated ethical dilemmas.

Curricular Approaches

While most plain-language communicators are practitioners and not academics, those who do teach in colleges or universities can benefit from others' approaches. Sawyer (1988) provides a multistep, almost syllogistic process for arguing about ethical issues. Possin (1991) critiques Sawyer's model and advocates a shorter, three-step process of arguing from analogy. Both authors, however, leave the definition of ethics to the reader. Hall and Nelson (1987) write that professional writing instructors might hesitate to discuss ethical considerations out of fear of violating students' expectations, fear of appearing dogmatic, fear of lacking adequate knowledge on ethics, and uncertainty on how to teach ethics. Nevertheless, ethics are important for students who will someday work in the public eye. Hall and Nelson (1987) advocate teaching a utilitarian perspective on the consequences of ethical decisions. They suggest creating fault trees or flow diagrams that identify consequences of decisions. Sturges (1992) also provides a flow diagram to help readers decide whether a particular action is ethical. Dombrowski (1995), among others, holds that ethics cannot be "technologized" into a flow diagram, a set of procedures, or any other reductive form; rather, writers must debate ethical questions with people who have knowledge of each situation's characteristics.

Rentz and Debs (1987) write about teaching ethics through analyzing sales letters. They also provide a striking example of a writer facing a difficult ethical situation. This writer worked for a company looking to expand a nearby airport, and she interviewed residents who would be affected by the expansion. When she could not reconcile the company's values with the residents' concerns, she resigned. Sims (1993) describes two cases that can help students understand the challenges of communicating honestly and truthfully and of navigating the interests and rights of their employers, the public, and themselves. Scott (1995) outlines a sophistic approach to teaching ethics. Scott focuses on the Greek concept of *nomos*, wherein ethics are relative and adapt in response to specific situations.

English department faculty, especially those with backgrounds in literary criticism, often find their personal values conflicting with the commercial and technical interests of the professions their technical writing students will enter (Russell 1993). Russell discusses a course introduced at the Massachusetts Institute of Technology in the 1920s, Engineering Publicity, as an attempt to teach students about the ethos and ethics of writing in the engineering profession. Russell suggests that if faculty in professional writing learn more about the history, curricula, and discussions of ethics in the professions their students seek to enter, they can help students become more responsible professionals and "more humane participants" in public discourse (107).

Hawthorne (2001) describes how the process of creating a code of ethics for their academic program helped students put abstract concepts of ethics into practice. Sullivan and Martin (2001) use narrative theory and the concept of the *apologia* to advance a narrative understanding of ethical issues: "When faced with an ethical choice, a person simply asks, 'what story will I tell about it when called to give an account before a jury consisting of all affected in any way?'" (269). Longaker (2005) combines the *paideia* tradition of instruction in rhetoric with the historic materialist perspective focusing on economic issues and class formation. Although he notes that students and other instructors may be wary of this approach, Longaker's goal is to critically engage students and attempt to shape the civic virtue of the new ruling class.

Articles from a 2004 special issue of *IEEE Transactions on Professional Communication* describe how instructors can use recent corporate cases to teach ethics. Each case provides numerous resources for understanding its important issues, thus avoiding the oversimplified either-or, all-or-nothing discussions of ethical behavior against which Porter (1993b) and Johnson (1998) advise. House, Watt, and Williams (2004) investigate rhetorical techniques used by Enron whistleblower Sherron Watkins. Zoetewey and Staggers (2004) show how stakeholder ethics, visual rhetoric, and risk communication intertwine in the case of an Air Midwest flight's fatal crash in 2003. Nelson-Burns (2004) describes the case of an eroded reactor head at the Davis-Besse nuclear power plant in Ohio. The case requires students to identify ethical issues related to withholding and revealing sensitive information. Strother (2004) uses the press releases issued by American Airlines and United Airlines on September 11, 2001, to show varying degrees of ethical consideration in an especially challenging corporate-communication scenario.

Other authors provide perspectives on semester-long courses. Ballentine (2008) describes a cross-curricular engineering communication course that brings together themes of globalization, intellectual property, design, and ethics. Kienzler and David (2003) describe a semester-long approach to integrating ethics into writing classes. Riley, Davis, Cox Jackson, and Maciukenas (2009) describe an approach to communicating engineering ethics through micro-insertion. Rather than teaching ethics in a freestanding course, instructors can use short cases with ethical components as class assignments or homework.

A series of articles from Markel outlines ideas on technical communication ethics that he develops more fully in his ethics book (2001). In the first article, Markel (1991) provides basic definitions of ethics and applies the literature of business and professional ethics to technical communicators and other professionals who communicate. Markel shows that communicators on the job must think through the conflicts of obligations to their employers, the public, and the environment. In the second article, Markel (1993) critiques the limitations of utilitarian or teleological approaches to ethics that focus on ends or consequences. Markel points to the second formulation of Immanuel Kant's categorical imperative—in your actions, always treat humanity as ends and not as means—as a rational, superior choice over the contingencies of utilitarianism. An extension of Kantian ethics found in John Rawls's description of justice in the ideal society is also relevant. Later, Markel (1997) discusses foundational, deductive approaches to ethics. Markel notes that we run the risk of oversimplifying these complex theories when we attempt to discuss them in accessible ways. Nonfoundational approaches to ethics can supplement the foundational, but they do not apply as broadly as foundational approaches (285).

Critical Theory and Ethics

As Dombrowski (2007) notes, the area of critical theory and ethics in technical communication has generated important discussions. Sullivan (1990) describes the tensions he sees between the need to teach students how to effectively write in workplace genres and the need to empower students to challenge political and ethical situations in corporate work. Katz's (1992) article, "The Ethic of Expediency: Classical Rhetoric, Technology, and the Holocaust," uses a memo written by a Nazi soldier during the Holocaust to discuss the ethic of expediency. Katz shows that this common workplace value can, if unchecked, override other humanistic concerns when people deliberate about important issues. In a companion article, Katz (1993) uses Berlin's concept of social-epistemic rhetoric to examine Hitler's startlingly effective propaganda campaign. In social-epistemic rhetoric, rhetoric functions within an ideology. Using Aristotelian models, Katz explains how Hitler manipulated deliberative rhetoric to redefine *praxis* and *phronesis* (social action and reasoning about ends and means) to empower his twisted ideology in Nazi Germany. More recently, Ward (2010) reviews a 1935 Nazi racial-education poster. Ward's analysis reveals the usefulness of both foundational approaches to ethics and postmodern nonfoundational approaches. Schroll (1995) applies principles from Katz, Habermas, and others in discussing ethical impacts of caller-identification telephone technology. Schroll urges technical communicators to be aware of ways that technology can inhibit ethical communication practices.

Porter (1993b) questions the value of academic discussions of technical communication ethics that portray workplace writers as solitary individuals rather than members of teams in larger organizations. Cases such as the Three Mile Island nuclear disaster and the launch of the space shuttle *Challenger* "sometimes portray the ethical decision as an either-or proposition: Either you cooperate with the company (and hence risk behaving unethically), or you act like a hero and blow the whistle (and risk losing your job)" (130). Porter worries that a focus on dramatic, large-scale cases tells students that the infrequency of such events means that students may never have to worry about ethics and that being ethical means "opposing the company when you know you are right" (131). Porter argues that instructors should include relevant laws and policies in discussions of writing corporate documents; they are parts of the rhetorical situation that writers often overlook. In corporate composing among multiple writers and reviewers, ethical responsibility "is socially constructed and therefore must be shared" (132).

Dragga (1997) interviewed 48 technical communicators to understand how they analyzed ethical situations. He cites philosopher Henri Bergson, who says that individuals' morality will be displayed in their responses to two main questions: "What will I do?" and "Who will I be?" (165). Dragga found that many of his respondents looked primarily to their employers for ethical guidance, that many followed intuition and feelings in ethical situations, and that one-quarter of them could not name a person they admired as a model of ethical behavior. In response, Dragga recommends a narrative approach to ethics that gives students stories from which they can draw, that identifies heroes in the profession exhibiting ethical behavior, and that allows instructors to share their ethical values and decision-making processes with their students (174–75). Faber (1999) critiques the role of intuition in ethical reasoning. Faber uses terms from social theory to identify some of the forces that shape our intuitive responses. These terms include Bourdieu's *habitus*, referring to habitual, mundane routines; Giddens's routinization, describing familiar patterns that people follow without critical reflection; and Fairclough's naturalization, describing how regularized uses of language perpetuate ideological and political meanings (1999, 192). Faber (1999) links these three terms to the work of Foucault, whose ethical work focuses on the "ways individuals unconsciously allow themselves to be transformed into organizational subjects" (195). Faber argues for critiques of intuition that "denaturalize" the agendas accompanying common actions and thoughts (200).

Feminist Approaches

Feminist approaches in many fields, including rhetoric, offer alternatives to traditional, masculine approaches that tend to embody rational thinking, focus on a single interpretation of truth, exclude emotional points of view, and accept data only from sanctioned sources. Dombrowski (2000a) identifies and summarizes several in his discussion of feminist ethics and the ethic of care. I will not duplicate Dombrowski's summary, but I will identify salient points. Feminist perspectives complement principles of dialogic communication, which I will discuss later in this chapter.

Lay (1994) identifies six key characteristics of feminist theory and argues for a redefinition of technical communication. These characteristics are celebration of difference, theory activating social change, acknowledgment of scholars' backgrounds and values, inclusion of women's experiences, study of gaps and silences in traditional scholarship, and new sources of knowledge (142). Lay identifies three actions that redefine technical communication in feminist perspectives. The first is to confront the myth of scientific objectivity and to understand the social construction of scientific knowledge. The second is to conduct ethnographic studies to help explicate the social and interdisciplinary natures of technical communication. The third is to study collaborative writing practices. Lay points particularly to possibilities brought to light in work on collaborative writing by Ede and Lunsford (1990). Lay notes that female technical writers whom Ede and Lunsford interviewed used a dialogic model of collaborative writing. The female writers viewed the multitude of voices and roles they took on as beneficial, while traditional, male-oriented processes regarded the multiple voices and roles as problems to overcome (Lay 1994, 154). Effective plain-language communication involves dialogue with audiences, not a monologue delivered at them.

Sauer (1993) reviews postaccident investigation reports from the mining industry. She writes that the public discourse of mining accidents privileges the rational (male) objective voice and silences human suffering, that officials exclude women's experiential knowledge (of things such as their husbands' health or the amount of coal dust on their husbands' work clothes), and that the conventions of mine accident discourse perpetuate silent power structures and exclude alternative points of view. Interpretation strategies that do not question unstated assumptions about gender, power, authority, and expertise "seriously compromise the health, safety, and lives of miners and, in a broader sense, of all those who are dependent on technology for their personal safety" (66). Sauer provides a list of improvements to mine safety suggested by women whose husbands died in a Kentucky coal mine explosion. But officials did not heed this list because it came from outside the mining power structure; the women had no voice in the discourse. Sauer writes that feminism forces us to acknowledge the moral, ethical, and political power of language and that a shift to a feminist perspective can reveal inadequacies in objective scientific discourse (78–79). Because BUROC situations so often involve an imbalance of power between organizations and their constituents, feminist perspectives are valuable for plain-language communicators.

In the introduction to their special issue of *Technical Communication Quarterly* on gender and technical communication, LaDuc and Goldrick-Jones (1994) write that gender research and feminist scholarship provide relevant and powerful means to understand situations in which professional communicators must consider social action, ethical responsibility, and civic responsibility. The articles in their special issue provide examples of women-centered scholarship that focus on women's contributions to the field, feminist methodologies that accommodate gender as a phenomenon, and postmodern feminist scholarship that critiques power differentials and the ways discourse constructs and perpetuates them. Further discussion of gender in technical communication appears in LaDuc (1997). "Looking through the lens of gender allows the writer-scholar-teacher to see how a number of communicative practices can be oppressive. More positively, feminist perspectives can also transform 'impersonal' questions of science, technology, and information access into issues that are personal, politically significant, and open to scrutiny and public debate" (LaDuc and Goldrick-Jones

1994, 247). In many BUROC situations, those who need information receive impersonal treatment and must struggle with technical information that they cannot understand well. On the other hand, plain-language communicators treat individuals in BUROC situations as people, not as problems, and help individuals respond to BUROC situations effectively.

Buber's Dialogic Ethics in the TPC Literature

Three recent articles and chapters by Katz and Rhodes (2010), Salvo (2001), and Dragga (2011) draw on the dialogic ethics of Martin Buber to establish ethical relationships between writers and their audiences. As these works share a foundation on Buber's dialogic ethics, they also provide specific examples of how to realize principles of dialogue in practice.

Buber was an Austrian-born, Jewish philosopher whose work appeals to a variety of audiences, secular and religious. Those who describe the dialogic view of ethics often cite Buber's *I and Thou* (1970). It is important to note that Buber wrote about dialogic ethics in many of his works and that he sought to apply dialogic ethics in his life. Buber wrote and taught in Germany in the early twentieth century; he moved to Jerusalem after the Nazis rose to power. In the latter part of his life, he lived among the *kibbutzim* social-collective communities in Israel. He lived out dialogic ethics in advocating for the presence of both Arabs and Jews in Israel (Friedman 1955, 8).

In *I and Thou* (1970), Buber describes two relationships one can have with others. In I–It relationships, one person speaks down to the other in technical dialogue; there is no true relationship between them. In I–You relationships (sometimes translated I–Thou), each stands in relation to the other; the relationship is reciprocal: I and You act on each other, and each reifies the other. While not every relationship is I–You, and relationships will not always stay in the I–You state, the I–You relationship is ideal.

Buber's depiction of the "narrow ridge" frequently appears in discussions of dialogic ethics. Buber describes the narrow ridge in *Between Man and Man* (1965). The narrow ridge is a place between two sides of an argument where the parties can meet if they regard each other as Thou and not It. As Arnett (1986) describes it, "The narrow ridge is a philosophical stance that undergirds behavior" (38). Two parties may separate because of significant differences: ideological, religious, philosophical. They may separate because of what Buber called existential mistrust—lingering suspicions about the other's true motives (49). Similarly, in "Hope for This Hour," Buber (1967) describes the human world as split into two camps, each of which thinks it embodies truth while the other embodies falsehood. Each side believes it has ideas while the other has only ideologies. Mistrust incites the two camps against each other (Arnett 2004, 79). In BUROC situations, an organization's constituents often feel like they must face off against the bureaucracy. Feelings of separation and distance from decision makers often coincide with physical separations between the groups. Plain-language communicators have frequent opportunities to create narrow ridges between organizations and their audiences.

The narrow ridge is a place from which people in a dialogue genuinely listen to each other and remain open to the other's persuasion. Participating in a narrow-ridge dialogue does not mean that the parties must compromise or that they must relax their convictions. It does mean that each party respects the other, demonstrates goodwill, and seeks to share community together. As Friedman (1955) writes, "Buber's 'narrow ridge' is no 'happy middle' which ignores the reality of paradox and contradiction in order to escape from the suffering they produce. It is rather a paradoxical unity of what one usually understands only as alternatives—I and Thou, love and justice . . . unity and duality" (3). Plain-language communicators create narrow ridges for their audiences by using the audience's language, respecting the audience's levels of literacy and understanding, and testing documents with members of the audience.

Ethics of Engagement through User-Centered Design

Salvo (2001) explores the ethical implications of user-centered design through Buber's approach to dialogue. User-centered design marks a shift in the philosophy of design for software and other technologies. Traditional approaches to design have separated designers and users. As such, designers are the experts while users might occasionally participate in usability testing during final stages of design. User-centered design practices, however, allow users to collaborate with designers in the design of a product. Salvo writes that effective collaborative design methods require "meaningful communication between users and designers, and dialogic ethics can guide the development of effective and humane technological design methods" (274). While perfunctory usability testing might allow a company to increase a product's efficiency or to claim user advocacy in their marketing campaigns, truly user-centered design practices empower users. Salvo (2001) describes a dialogic approach as an ethical, humane counter to the ethic of expediency (Katz 1992) prevalent in many workplaces: "Aligning dialogic rhetoric with usability creates a background for understanding usability as ethical design praxis rather than an efficient mode of technological design and manufacture" (Salvo 2001, 276). Salvo provides three examples of dialogic, user-centered design practice.

Salvo first describes Pelle Ehn's participatory design approach, an approach sometimes called Scandinavian design. The social context for participatory design in Scandinavian countries differs noticeably from that in the US. Scandinavian workers tend to belong to unions, and unions wield significant power; laws and agreements often regulate negotiations between workers and management; the workplace there is more democratic than in the US (Ehn 1993). Ehn (1993) discusses large projects in Scandinavia in which management and labor, over months and years, used dialogue and collaboration to reach consensus. Although the American perspective on relationships between management and labor differs greatly from that in Scandinavia, Salvo (2001) notes salient principles of dialogic engagement that do apply: when workers experience discomfort of any kind when using a product, product designers need to know; factors contributing to discomfort may well be social, organizational, and cultural, not merely technical; and political forces may silence the voices of workers and technology users, but active advocacy through dialogue can effect positive changes (278–81).

Salvo's (2001) second example is the process for designing tactile signage for blind and sight-impaired individuals used by Robert Whitehouse and his team. Salvo writes that user-centered design does not apply the traditional scientific process to reach generalizable conclusions; rather, it is a postmodern process like those described by Sullivan and Porter (1997), responding to local conditions and concerns. Whitehouse (1999) describes how his team developed signage to help individuals find the Lighthouse, an office serving the blind and the sight impaired. The users of these signs were diverse: some were born blind while some lost their sight later in life; some could read Braille, but many did not; some were older, some younger. Whitehouse's team created prototype signs and then tested them extensively with their target users. The resulting product works well for particular users in a specific community. Salvo (2001) writes, "The distinguishing characteristic of dialogic methods of design and what sets them apart from scientific methods is reliance on information provided in the form of feedback. It is not scientific, not neat or particularly efficient in the design process, but it makes for much improved use of the products being designed" (284).

Salvo's (2001) third example is the development of TOPIC, an online system for writing instruction created at Texas Tech University. Writing from his experience, Salvo describes the many challenges of creating an online system that could support composition instruction at a large state university and perform effectively for a broad array of users: undergraduate students, graduate student instructors, other part-time instructors, tenure-line faculty, and writing program administrators. After beginning with surveys of instructors and administrators, the design team created an early version of TOPIC. Classroom teachers then experimented with the system, and their suggestions led to changes in future iterations of the system. This dialogic process was recursive and ongoing: "User-participatory design, while related to user-centered design, puts users in much closer contact with designers and often blurs the boundaries between feedback, usability, and design. Changes users suggested in the feedback process often quickly became new design elements, making the distinction between feedback and new information blurry, uncomfortably blurry at times" (286–87). Salvo writes that although traditional software development methods might have produced a system like TOPIC more efficiently, the involvement users at

Texas Tech had in the TOPIC design process helped make TOPIC a better tool
for them:

> In many ways, the interaction resulted in greater understanding of the
> writing program at Texas Tech. This system view of the composition pro-
> gram, as understood by participants at very different phases of its activity,
> is as valuable to the department as the software product has been, giving
> additional value to the process. That process was guided by ethical prin-
> ciples of dialogic engagement. (287)

Participants in Buber's I–You relationship understand their mutual depen-
dence. Salvo (2001) writes that a dialogic, ethical approach in participatory,
user-centered environments allows stakeholders to rearticulate their roles continually. "In developing new design models, technical communicators have an
ethical responsibility to help users become more informed users while making
producers more responsive producers, and the clearest route to this goal is to
raise the interaction, the dialogic interaction, between these populations" (289).
Plain-language communicators, who frequently incorporate testing with users in
their work (e.g., Cutts 2009), do ethical work when incorporating user feedback
into the documents they develop.

Dialogue in Ethical Frames of Technology

Using a case study on email use in an organization, Katz and Rhodes (2010)
examine the ethics of digital communication. They seek to reveal ethical facets
of "technical relations" that may ultimately enable readers to move beyond limited, compartmentalized, and even conflicting ethical perspectives in the world
of business and industry. They discuss ethical frames inherent in the evolution of
communication technologies, which they say the technical communication field
has been slow to recognize. Dynamic and socially constructed, ethical frames are
"a set of philosophical assumptions, ideological perceptions, and normative values
underlying and/or guiding how people relate to and exist with technology" (231).

Katz and Rhodes (2010) describe six frames in Platonic and Aristotelian terms
(232–41). They begin with the false frame, which is no frame. Plato separated
meaning (*episteme*) from art (*techne*); in the false frame, technology contributes
nothing of value. Plato held that the technology of writing was false knowledge,
an imitation of knowledge. Technology in this false frame might provide entertainment or an indulgence. In the first frame, the tool frame, technology is a
means to an end. Here, ethics focus on use of the tool or craftsmanship. In the
second frame, technology is both means and end. The primary ethical issue is to
determine whether the technological end justifies the technological means. For
example, the question of whether to upgrade a web server to serve additional
traffic involves both ends and means. In the third frame, the autonomous frame,

technology produces its own means, ends, and moral codes (productivity, speed, efficiency). Societies that value material wealth see productivity, speed, and efficiency as complementary ethical values, and technology becomes an autonomous, self-contained ethical entity. In the autonomous frame, the values of technology can become values of the organization. In the fourth frame, the thought frame, technological strategic reasoning becomes the dominant mode of thought. Technology dictates organizational philosophies and ways employers and employees think. One ethical challenge in the thought frame is to continually recognize employees as unique people and not merely as technical skill sets. The fifth frame is the being frame. Katz and Rhodes use Heidegger's term of "enframing" to describe this technological way of knowing and being. All the other frames lead to means–end relations as our primary way of being in the world. In the being frame, our consciousness is a kind of rational ordering that begins in means–end relations; documents and people (human resources) are objects in reserve and ready for use. People connect to the workplace through cellular phones and electronic networks. Through coercion or by choice, they often stay connected to the workplace from home and even vacation through electronic devices.

After identifying these five frames focused on means–end relationships between people and technology, Katz and Rhodes (2010) propose a sixth frame, the sanctity frame, encompassing all the others. In describing the sanctity frame, they use Buber's concepts of I–Thou and I–It. While I–It relationships are based on strategy and means-end relations, I–Thou relationships are based on mutual respect. "In this frame, both humans and technology (often merged) would no longer be an 'It,' a 'thing' completely objectified, a means to end, and thus rendered as a standing-reserve. Rather, outside technical consciousness, an 'It' might become a 'You,' a subject that is intimately related to 'I'" (250). Katz and Rhodes imagine that in the frame of sanctity, employers might see their employees not only as Its able to do work but as Thous worthy of respect and capable of providing reciprocal respect through their technology use and their interpersonal interactions (251). While Katz and Rhodes do not label this sanctity frame as a dialogic, ethical stance, it certainly is dialogic as they describe it.

Lessons from Dialogic Codes of Conduct

After scandals at corporations such as Enron and WorldCom, codes of conduct have received renewed attention. In response to these corporate implosions, the US Securities and Exchange Commission, the New York Stock Exchange, and NASDAQ have updated their requirements for companies to adopt codes of conduct (Dragga 2011). Corporate codes of conduct "operate in a legal and ethical environment, addressing multiple audiences both inside and outside the organization" (Dragga 2011, 5). Dragga says individuals must navigate between four sectors of activity: ethical and legal, unethical and illegal, legal but unethical, and ethical but illegal.

As in an earlier article, Dragga (1997) applies Henri Bergson's theory of ethics to understand the environment in which corporate employees make ethical decisions. As Dragga describes Bergson's theory, shared perspectives in a society (e.g., laws, rules, regulations) push individuals toward beneficial behavior and stability. The ethical-and-legal and unethical-and-illegal sectors are associated with this push. Morality can pull individuals toward higher callings and aspirations, leading toward societal progress. Heroic individuals challenge existing perspectives when they respond to the pull into the ethical-but-illegal and the legal-but-unethical sectors. Corporate codes of conduct typically emphasize the push of regulations and not the pull of morality (6–7).

Dragga (2011) uses Buber's description of the dialogic relationship to explain how codes of conduct can inspire ethical action. He applies Buber's principles this way: "To be ethical is to treat each other as a 'thou' (a human being) instead of as a simple 'it' (a thing to be utilized). The I-thou relationship encourages dialogue, while the I-it relationship allows only monologue (the authoritative I and the submissive it). In the I-thou relationship, mutual respect and genuine reciprocity of action are possible" (7).

Using rankings developed by the Ethisphere Institute, an organization providing training in business ethics and corporate social responsibility, Dragga (2011) analyzes codes of conduct for three companies then identified as highly ethical: Verizon Wireless, Granite Construction, and BP. His analysis focuses on five factors he proposes as contributors to a dynamic I–Thou relationship in creating dialogic codes of conduct: language emphasizing cooperation versus compliance, identification of code authors, clarity of ethical versus legal obligations, inclusion of humanizing pictorial images, and coverage of heroic individuals (8).

The three codes of conduct do not provide many examples of a cooperative, dialogic I–Thou relationship in the three codes. Verizon Wireless starts its code with some "we" statements implying collaboration and shared goals among management and employees. Later, however, Verizon shifts to a heavy-handed tone emphasizing compliance. Questions and answers appear in the document margins, in an apparent attempt to mimic dialogue between employees and management. The company has all the answers and the workers have none; Dragga (2011) calls this a "monologue disguised as a dialogue" (10). While he praises the promising start to Verizon's code of conduct, Dragga finds that the code "ultimately sacrifices cooperation for compliance and chooses a monologic I-it relationship in which it is permissible to ask questions but inconceivable to question the guidelines or participate in their writing or revision" (11).

Granite Construction's code is more successful in maintaining a dialogic stance. Unlike Verizon, Granite includes pictures to humanize its message—including one of the company president and his dog. Statements of core values depict a collaborative ethos. The code is regularly reviewed by a team of Granite employees, although these employees are not named. The code also identifies heroic individuals embodying ethical actions: company "founding fathers" Walter Wilkinson and Bert Scott. Although the code has 14 pages of compliance guidelines compared to two pages of core values, the use of "we" throughout helps emphasize collaboration. The code concludes, however, with statements emphasizing employee compliance and respect for the corporation. Overall, Dragga (2011) finds that Granite tries to build both "a collaborative and dialogic environment and the foundations of a genuine I–Thou relationship, but sporadically sacrifices the spirit of cooperation to the language of compliance" (12).

The code from BP likewise shows some signs of dialogic relationship with readers before eventually shifting back to emphasizing compliance. The CEO's introduction does emphasize collaboration and equality, and it invites candid conversation (Dragga 2011, 12–13). Although BP's code uses images, the images do not humanize the code effectively. The CEO appears in a photograph while subordinate workers appear more abstractly in drawings. The code does not depict heroic individuals who inspire others toward ethical action, nor does it effectively separate matters governed by law from matters governed by BP policies. Similar to the code from Verizon, much of the language in BP's code is monologic and focuses on employee compliance through an I–It stance (Dragga 2011, 13).

Dialogic Communication Ethics: A Model for Understanding Ethical Implications of Using Plain Language in Technical Content

The professional strand of TPC literature reveals strong convictions that technical communicators should strive for clarity in the documents they create. Clarity—often the primary reason for using plain language—is an important ideal because unclear messages might mislead readers. An unethical person or company could use unclear or misleading messages to take advantage of readers, and readers might make unwise decisions if a message misleads them. Many agree with Kostelnick (2008) when he writes that data designers—and, by extension, other technical communicators—have a moral imperative to create clear documents, even though Kostelnick does not explain from whence that imperative comes.

Through foundational ethical approaches and dialogic ethics, we can better understand the technical communicator's imperative to be clear—which communicators may enact through plain language. First, I will review Clark's (1987) third approach to technical communication ethics, which he calls rhetorical, to show how it applies to clarity and plain language.

Critiquing Clark's Rhetorical Approach to Ethics

Clark (1987) hints at a dialogic approach to ethics at the end of his article on professional and academic ethical perspectives. After describing the limitations of the *professional* and *academic* points of view, he proposes a third that he calls *rhetorical.* Clark points to Plato's *Gorgias* and *Phaedrus* and Aristotle's *Rhetoric* as texts that describe examples of collaborative and cooperative ethical communication. Clark (1987) argues for a cooperative approach to ethics in technical communication that centers on "the well-being of all the people who must interact in the process of sharing technical information, the one community to which all participants in the process of communication belong and are responsible.... This ethics makes communication both responsible and practical—properties which cannot be ethically separated" (194). While direct interaction with audience members is most helpful to technical communicators but is often impossible, Clark says it is ethical to anticipate and imagine what audience members might say in a dialogue with writers of, say, a manual. In short, Clark advocates taking a dialogic, ethical stance toward the audience, with a goal of reaching consensus. Clark postulates that better collaboration between the managers overseeing the launch of the space shuttle *Challenger* and the engineers who sought to dissuade them could have ensured fully shared meaning between the groups and thus prevented the tragic loss.

A set of articles critiquing Clark's (1987) views on ethics appears in an issue of the *Journal of Computer Documentation.* Davis (1994) says Clark's article is not as much about ethics as it is about defining the competence of technical communicators. Davis finds Clark's definition of ethics overbroad and his views of collaboration and agreement in the workplace naïve, and he questions Clark's selection of the *Challenger* disaster as a means to envision ethical communication (16–18). Walzer (1994) complains that Clark misreads Plato and Aristotle. While Plato does describe dialectic as a collaborative means of finding truth, Plato holds that truth is revealed in a "'eureka' experience" and not through social construction (Walzer 1994, 28). Aristotle, Walzer writes, does not see rhetoric as "an art that necessarily fosters a cooperative, participatory process. . . . The courtroom, the political assembly, and the podium on ceremonial occasions have institutional purposes that would not allow for, much less invite, the collaborative exploration that Clark claims can be derived from the *Rhetoric*" (29). Nevertheless, Walzer (1994) does credit Clark for explaining why professionals and academics in technical communication "talk past one another" when discussing ethics (31). Brockmann (1994) calls Clark's approach "flimsy" but quickly concedes that "Clark's focus upon the need to depend upon each other's goodwill in the area of ethical technical communication is correct . . . and is, perhaps, our only recourse" (11; ellipsis in original) T. R. Girill (1994) notes that in technical communication, often there is no real community, no real cooperation, and no common project to unite writers and their audiences (21). While Girill writes that Clark's rhetorical ethical approach applies only to situations in which parties cooperate overtly, Girill suggests that negotiations based on the parties' shared interests provide a more robust ethical approach (24).

In the same issue of the *Journal of Computer Documentation,* Clark (1994) revises his approach to technical communication ethics. In this essay, Clark emphasizes the value of dialogue within a community, exemplified in the critiques of his article. Clark agrees with Davis (1994) that professional ethics—the special obligations held by members of a profession beyond those imposed by law, market, and morality (17)—is the true subject of his original article. Clark (1994) writes that the exchange among the four other authors and himself in the *Journal of Computer Documentation*, where the authors write as listeners, embodies rhetorical ethics (36).

To state his revised view of ethics, Clark (1994) writes that communities hold together by *consent*, not *consensus* as he had previously argued (1987); he directly discusses the value of argument and disagreement to the social processes of knowledge construction. Consenting communities within a profession "are maintained through conflicts that function continually to reconstruct that consent. . . . Ethical communication in this complicated context must function in an exchange that enables the limitations of any perspective to be extended by the differing and conflicting insights of others" (Clark 1994, 38). For Clark, then, dialogue is crucial for enacting ethical behavior. Clark also makes this point in his book about the potential of rhetoric and composition in democratic education, *Dialogue, Dialectic, and Conversation* (1990).

Although Clark (1994) does not specifically address the ethics of communicating clearly or of the plain-language movement, we may infer that it is ethical to use language that promotes dialogue between parties. While Girill (1994) writes that technical communicators often have little contact and little sense of community with their audience members, plain-language communicators are likely to meet with audience members more often. Through usability testing, outreach to individuals with low literacy, and other dialogic interactions with members of target audiences, plain-language professionals can act ethically.

Applying Foundational Approaches and Dialogic Ethics to Understand the Ethical Value of Plain Language

Foundational concepts of ethics support the imperative for technical communicators to create clear messages. Utilitarianism reinforces the practical value of communicating clearly; however, excessive reliance on utility could lead to a damaging enshrinement of the ethic of expediency (Katz 1992). Through Kantian ethics, we appreciate the rights that audiences have to experience life as ends and not as means to an end. When communicators provide audiences information that best informs them, they act ethically. We can understand communicators' obligations in light of audiences' rights. The concept of care gives technical communicators another way to examine—and reaffirm—the relationships between their companies and their audiences and to consider the many ways in which those relationships manifest. Through feminist perspectives, communicators can acknowledge and address the imbalance of power often present between bureaucracies and their constituencies: constituents often have little power, the bureaucracies often have much, and plain language can help address the differential.

Through dialogic ethics and the ideal of the I–You relationship, the importance of clarity becomes paramount. The dialogic approach requires rhetors to view the audience not merely as important, but as essential to their own being. When clarity supports mutual understanding between an organization and its audience, it supports the full potential of the I–You relationship. The work of Levinas complements Buber's by emphasizing respect for the audience, the Other. Often, mistrust and misunderstanding separate the rhetor, or the "I," from the Other significantly. Like Buber, Levinas finds the root of ethics not in abstract principles but in our encounters with others, or the Other. In Levinas's view, ethics is not an abstract system of principles or a sense of duty; ethics comes from awareness in our relationships with others. We can't judge what will be ethical conduct toward another until we communicate reciprocally with him or her (Dombrowski 2000a, 71). Through reciprocity, we learn to treat the Other as You and not It.

Buber's approach to dialogue is one among several. Another approach frequently applied in discussions of rhetoric and composition is that of Mikhail Bakhtin. Mendelson (1993) writes that Bakhtin's rhetorical theories start from the notion that all discourse is within a matrix of chronological and situational contexts and that all discourse is part of an ongoing discussion; writing or speaking, we are always in dialogue with others (295). Participants in a dialogue do not alternate but participate simultaneously in each utterance. Bakhtin's concept of addressivity reinforces the idea that speeches (and texts) address specific audiences. Addressivity describes not only the familiar patterns in a dialogue between speakers but also an internal process of intralocution in which writers answer and anticipate the response of the dialogical partner (297). While addressivity helps rhetors look outward toward partners in dialogue, Bakhtin's concept of heteroglossia turns inward and focuses on the dialogue that exists within the nature of language itself. Writers "use language that retains the echoes of others who have used it before. Even in the isolated act of composition, we are in dialogue" (Mendelson 1993, 298). Clark (1990) provides another introduction to Bakhtin (8–18).

As Salvo (2001) describes them, Bakhtin focuses on linguistics whereas Levinas focuses on identity and the need to recognize the humanity of the Other (as does his mentor Buber, by extension) (275–76). While Bakhtin's principles are relevant to a contemporary discussion of ethical dialogic communication, Buber's concepts of I–It relationships, I–You relationships, and the narrow ridge are especially salient for BUROC communication situations. Buber's concepts illuminate the divide between a bureaucracy and its audiences. The ideal described in depictions of the I–You relationship gives writers a model that can influence their communications for the better.

Conclusion: Using Plain Language to Apply Dialogic Communication Ethics

In BUROC communication situations, described in Chapter 1, metaphorical and physical distances separate bureaucratic organizations and their constituents. Military veterans returning from service often live far from Veterans Administration offices and facilities, but they have rights to use VA services. Medical patients who need even routine care have rights to access medical treatment, but the ultimate arbiters of the coverage they will receive are distant from them physically and figuratively, shrouded by voluminous, complicated contracts and policies. Consumers and citizens often feel that they are in one camp while the bureaucracies are in another. Although the two sides need and depend on each other, they may feel like opponents in a grueling battle.

The BUROC framework helps writers, often working within or for bureaucracies, to understand the situations audience members face. Bureaucracies are often complex and confusing; plain-language communicators can provide direction and guidance. Bureaucratic jargon is unfamiliar to those on the outside; writers can use plain language to explain important concepts and provide examples. Audience members have rights, and the ethical tradition shows the importance of respecting them. Using plain language, writers can help individuals better understand their rights in order to use them well.

Feminist perspectives recognize that power differentials typically favor bureaucracies over citizens, making it especially important for individuals to use their rights fully. Critical situations require citizens to interact with bureaucracies—insurance companies, hospitals, courts of law, and so on. Using the BUROC framework, writers can recognize how critical and important certain events can be. Citizens and consumers in the audience are often the Other to those inside a bureaucracy, but writers can see them as partners in dialogue, as Yous and not Its.

By electing to use plain language when it benefits the audience, a bureaucratic organization regards its constituents as You, not as It. Plain language supports dialogue when constituents need to act in situations that are atypical or unfamiliar to them and yet critically important. Bureaucratic organizations can enact Buber's concept of the narrow ridge by taking a dialogic, ethical stance and by using plain language to address feelings of mistrust or anxiety for their constituents.

Questions and Exercises

1. Find an ethics case dealing with technical communication. You might review one of the articles or books discussed in this chapter; STC's *Intercom* magazine is another place to look. Use some of the ethical principles described in this chapter to respond to the case in around 250 words.

2.   Which ethical principles, if any, do you typically associate with admonitions to write in plain language? Consider why you have these views. Find a source discussed in this chapter that appears to resonate with your views, and then read the original source. Summarize the source in a memo of around 300 words.

3.   Have you ever experienced a conflict in which you felt like you were on one side of a battlefield and your opponent was on the other? Were you able to find a way to meet with your opponent on a "narrow ridge" between you? If you did find the narrow ridge, write about 250 words about how you got there and what you accomplished there. If you could not find the narrow ridge, write about what you might have done differently to get to the narrow ridge.

3

PERSPECTIVES ON PLAIN LANGUAGE AND ETHICS FROM PROFESSIONALS AROUND THE WORLD

The previous chapter reviewed the literature on ethics in the fields of technical and professional communication (TPC) and of plain language, and five key principles emerged. First, the review showed that professionals consider it ethical to communicate to audiences clearly and unethical to confuse or mislead them. Second, it showed how effective communication, which plain-language communicators strive to provide, meets the ethical standard of utility or the pursuit of the greatest benefit; however, utility pursued to the extreme can lead to unethical behavior. Third, the review showed that the Kantian perspective on individuals' rights and on imperatives to respect those rights complements goals of plain-language communication. Fourth, the review showed how feminist perspectives on ethics and communication recognize and address imbalances of power; plain-language communicators routinely write for audiences who lack power and agency in the situations they face. Fifth, the review showed how Buber's dialogic ethics provides a means of understanding and theorizing plain-language communication.

In this chapter, I examine the ways that practitioners of plain language around the world define ethics, and I review the extent to which they view the practice of plain-language communication as having an ethical dimension. The plain-language communicators in this study tend to agree that ethics affect many aspects of plain-language use and that plain language can support ethical action. They recognize that a plain message is not necessarily ethical, and they disagree over whether audiences have a right to receive information in plain language.

I sought to stake out new territory in the literature on plain-language ethics. I wanted to understand how plain-language communicators define ethics, how they define their ethical responsibilities, whether they believe plain language is a civil right, whether they believe plain language provides a means of acting ethically, whether they believe that plain language supports Buber's (1965, 1970) views of dialogical ethics, and whether they believe the BUROC model effectively identifies opportunities for plain-language communication. I surveyed a diverse group of plain-language professionals for their views on ethics and plain language. While I cannot claim that their responses generalize to all plain-language professionals, themes that emerged from their responses provide a foundation for further discussions of ethics and plain language.

Plain-Language Professionals Participating in This Study

In the spring and summer of 2013, I recruited 24 active practitioners of plain language to respond to my questions about ethics and plain language. I used websites of plain-language organizations and social media to identify practitioners around the world. Table 3.1 lists the professionals in alphabetical order.

Overall, this group of professionals, 19 female and 5 male, has significant experience in plain-language work and is diverse geographically. The group represents eight different nationalities and a range of two to 35 years of experience working with plain language, averaging 18.3 years of experience. All participants completed at least a bachelor's degree; several have graduate or professional degrees as well. Common fields for their academic degrees include English, linguistics, education, and law.

Method

I used research methods in a protocol approved by the Boise State University Institutional Review Board. I created a questionnaire to collect information on the professionals' views of ethics. The questionnaire focused on professionals' definitions of ethics, on the sources of influence on their ethical views, on whether they believe that plain language is a civil right, and on whether plain language supports Buber's (1970) ideal of I–You communication. I contacted potential participants via email to recruit them for the study. Those who documented their informed consent to participate chose between responding to the questionnaire via email, telephone, or Skype; most preferred email. I provided the questionnaire in advance to those who preferred to speak with me via telephone or Skype. I analyzed all the responses in NVivo 10 software using principles of inductive thematic analysis described by Boyatzis (1998). This process includes identifying themes in data, comparing themes within the data, and applying codes to organize themes. The number of responses for each question varied; some participants chose to skip some questions.

TABLE 3.1 Plain-language professionals participating in this study.

Name	Professional activities
Carolyn Boccella Bagin	President of the Center for Clear Communication, a Maryland firm serving a variety of clients including the American Bar Association and the Internal Revenue Service.
Michelle Black	President of Simply Read Communications Group in Ontario, Canada, working with corporate, government, and nonprofit clients.
Deborah Bosley	Owner of The Plain Language Group in Charlotte, NC, working with corporations, government agencies, and banks. She is associate professor emerita of English at UNC Charlotte.
Peter Butt	Emeritus professor of law at the University of Sydney and a past president of Clarity International. His many books include *Modern Legal Drafting*.
Kathryn A. Catania	A plain-language communication specialist and cochair of the Plain Language Action and Information Network for US federal employees.
Annetta Cheek	Chair of the executive board of the Center for Plain Language. She worked in US federal agencies for more than 25 years and continues to advocate for plain language.
Martin Cutts	A longtime advocate for clear language in the UK. He is the director of an editorial and training business, Plain Language Commission. His books include the *Oxford Guide to Plain English*, now in its fourth edition.
William DuBay	Provides plain-language writing and training services and publishes the *Plain Language at Work Newsletter* through his consultancy, Impact Information, near the Puget Sound.
Sandra Fisher-Martins	Founder of Português Claro, a communication firm serving Portuguese audiences. The online video of her TED talk about plain Portuguese has more than 400,000 views.
Sarah Fox	A former contract lawyer in the UK now providing plain-language training for the construction industry through her company, Enjoy Legal Learning.
Caryn Gootkin	Owner of the editing firm In Other Words in Cape Town, South Africa, and a board member of The Big Issue, an NGO serving unemployed and marginalized adults.
Frances Gordon	An international plain-language consultant in Lisbon, Portugal. She previously worked on educational projects and corporate ventures in her native South Africa.

(Continued)

TABLE 3.1 (Continued)

Name	Professional activities
Simon Hertnon	Managing director of Nakedize, a New Zealand training and consultancy firm. He is the author of several books including *Clear, Concise, Compelling*.
Robert Linsky	An information-design practitioner for more than 25 years. He is director of information design at NEPS and serves on the board of Plain Language Association International.
Rachel McAlpine	An author in many genres and codirector of Contented Enterprises in New Zealand, providing services in writing, content strategy, and accessibility.
Christine Mowat	Founder of Wordsmith Associates, a communications firm serving clients across Canada, and author of *A Plain Language Handbook for Legal Writers*.
Karine Nicolay	Project coordinator for IC Clear, which is creating a postgraduate course in clear communication for European professionals. She lives in Belgium.
Karen Payton	President of Bright Communications in Ontario, Canada, providing plain-language training and writing services.
Audrey Riffenburgh	A senior health-literacy specialist for the University of New Mexico Hospitals and owner of a consultancy, Plain Language Works.
Write Limited	A clear-communication firm in New Zealand. Team members who responded collaboratively to my questions include Diana Burns, Lynda Harris, Rosemary Knight, Inez Romanos, and Colleen Trolove.

Why They Work with Plain Language

I asked the professionals why they chose to work with plain language. The factors driving their decisions range from the pragmatic to the personal.

Several cited pragmatic reasons why they chose to work in plain language. Four said that plain language helps them create better documents. Said Kathryn Catania, who works in government, "Producing well-designed documents using plain-language techniques such as informative headings, active voice, and common terminology will only increase the public's understanding of government requirements and services" (pers. comm., June 28, 2013). Four others found that plain language provided a logical step in their development as professionals. One of these, Carolyn Boccella Bagin, said, "My whole career has always involved communication in some aspect, so the move to plain language just evolved naturally.

The practice of plain language just crystallized much of what I already believed and saw practically applied in several of my early jobs" (pers. comm., May 30, 2013). Two professionals said that seeing important documents created in plain language inspired them. Annetta Cheek admired a plain-language regulation for a complex situation involving natural-gas royalties (pers. comm., June 10, 2013). Deborah Bosley found motivation in a PLAIN conference she attended (pers. comm., June 17, 2013).

Others cited personal experiences that link the pragmatic value of plain language with personal experiences and with opportunities to help others. Five wrote that working with people with low literacy led them toward plain language; their experiences include teaching English as a second language, working with immigrants, working with adult learners, and working with individuals with disabilities. Said Michelle Black, "Working in literacy as a volunteer and taking a literacy worker's course, I learned that there was a movement to 'translate' or 'interpret' my mother tongue [English] into another form of English. I was hooked from there" (pers. comm., June 10, 2013). Five respondents wrote that working in plain language is a way to help people. Karen Payton's comment will likely resonate with many plain-language practitioners; the situation she describes is not unique to her native Canada:

> Organizations and governments in Canada have been producing documents that require above-average English comprehension skills. Given Canada's literacy rates and ESL citizens there is an unfair expectation that people will understand. These organizations have an obligation to communicate clearly—governments, financial institutions, the health care industry, etc. For the most part, the onus is on the audience to understand, not on the organization to communicate clearly. We need to change that. (pers. comm., May 27, 2013)

Finally, five said they chose to work in plain language after struggling themselves to read and use confusing, unclear documents. Rachel McAlpine said that as a high-school teacher, the "river of jargon" she received from the Ministry of Education appalled her. "I was the target audience, yet I could barely understand the theory, let alone figure out what I was expected to do" (pers. comm., June 15, 2013). Sandra Fisher-Martins said, "Obscure language started getting on my nerves and, as there was no plain-language movement in Portugal, I decided to start one" (pers. comm., July 8, 2013).

The common thread among the professionals' reasons for working in plain language is the potential that plain language holds to make situations better for people who need to understand and respond to a message. The struggles people have with unclear, convoluted documents are real. Many of the professionals have seen how others have benefitted from using plain-language documents, and several have experienced those benefits themselves.

Defining Ethics

The questionnaire to which the professionals responded places related questions into sections; capital letters identify each section. For example, prompts A1 through A4 come from the same section of the questionnaire.

> *Prompt A1*: Ethics is often defined as a set of principles for determining good and right conduct. That said, many understand and apply ethics differently. How do you define "ethics" or "ethical behavior"?

The definitions professionals provided are similar to those discussed in the previous chapter's literature review. Eight professionals said that ethics involves doing the right thing; two of these added that ethical behavior is doing the right thing even without witnesses. Six stated that societies and communities shape ethics. Deborah Bosley said that ethical behavior is "socially agreed upon behavior that tends to benefit the most people" (pers. comm., June 17, 2013). Peter Butt said that ethics is a set of principles to guide people in their relations to each other and to society as a whole (pers. comm., July 3, 2013). Annetta Cheek (pers. comm., June 10, 2013) and Sandra Fisher-Martins (pers. comm., July 8, 2013) noted that individuals have ethical responsibilities to the environment as well as each other.

Four said ethical behavior involves being open, honest, and transparent with clients. Three said that ethics involves respecting others. As a Write Limited staffer wrote, "For me, ethical behavior is about three major principles: honesty, fairness, and respect. These principles apply in both the personal and professional spheres of life" (pers. comm., July 1, 2013). Three others said that ethical behavior helps others. Finally, Peter Butt referred to the Golden Rule, treating others as he would like to be treated (pers. comm., July 3, 2013).

The responses to this question overlap substantially with the professional perspectives on ethics described in chapter 2. Ethical behavior is also appropriate professional behavior (Radez 1989; Shimberg 1989b; Michaelson 1990; Markel 1991). It is ethical to communicate clearly and unethical to obfuscate, whether with words (Shimberg 1989a; Walzer 1989a) or with visuals and design (Herrington 1995; Bryan 1995; Allen 1996; Dragga 1996).

Defining Ethical Responsibilities for Plain-Language Communicators

> *Prompt A2*: Do you believe that a communicator in your line of work has ethical responsibilities? If so, please identify a few of the ethical responsibilities of writing in plain language.

The professionals overwhelmingly agreed that plain-language communicators have ethical responsibilities. Although the prompt refers only to *writing* in plain language, Audrey Riffenburgh and Carolyn Boccella Bagin rightly pointed out that plain-language communication involves audience analysis, visual design, testing with audience members—much more than writing alone.

Several said that ethics is essential to effective plain-language communication. Deborah Bosley said, "The whole field is about the ethics of—and maybe the lack of ethics—when people create information that others have to use but cannot understand. I think we're all about the ethics of language use" (pers. comm., June 17, 2013). Said Frances Gordon, "I think that plain language without ethics is pointless. I believe that an ethical view is what differentiates plain language from related disciplines" (pers. comm., May 21, 2013). Lawyer and legal writing expert Peter Butt wrote, "Deliberate ambiguity or complexity designed to cloud meaning, or put off people from reading the fine print, is contrary to the ethics of the plain language movement" (pers. comm., July 3, 2013).

Many said that honesty, clarity, and a commitment to the truth are part of ethical behavior for plain-language communicators. "As a writer and communications professional it's my obligation to present information clearly and truthfully," wrote Karen Payton (pers. comm., May 27, 2013). Caryn Gootkin said, "You need to stay true to the content, tone and style of the author while ensuring the text is accessible to a wide audience" (pers. comm., June 11, 2013). Audrey Riffenburgh notes that some challenges arise through seeking to write clearly: "At times, we may make meanings explicit which the authors meant to keep implicit. How we respond in that case, is an ethical issue" (pers. comm., June 25, 2013).

Six professionals said they saw an ethical responsibility to be sensitive to the audience of a message. Christine Mowat said, "I think that plain language respects the differences within our readers and adapts to levels of literacy, interest, background knowledge" (pers. comm., June 13, 2013). Michelle Black and Sandra Fisher-Martins feel ethically responsible to serve as user advocates—especially for those most vulnerable. The accessibility of a message is important as well. Rachel McAlpine eloquently summarized the relevance of accessibility:

> "Write for your reader" is an ethical injunction as well as a plain language guideline. This responsibility extends further today than ever before. Your reader may be blind or paralyzed; may be reading on a smart phone on a bus; may be from a different language background; may be listening to your words, not reading them. Your reader wants information for a particular use. And it's our job to satisfy these readers. (pers. comm., June 15, 2013)

Four professionals described how plain language can help readers exercise their rights to understand information affecting them. A Write Limited staffer wrote that ethical behavior "centers on people's needs and rights to understand information. It is unethical to expect people to behave according to information they can't easily understand. We [at Write] see our ethical responsibility as helping people to gain access to information that is important to them, by encouraging those who produce that information to write it more clearly" (pers. comm., July 1, 2013). Annetta Cheek, a longtime civil servant, said, "People have a right to understand what their government does. Therefore, government communicators have an obligation—and I guess it's an ethical obligation—to communicate clearly to the citizens to whom they are writing or to whom they are speaking" (pers. comm., June 10, 2013).

Three others said that plain language can help address social inequalities and injustices. Karine Nicolay has worked with immigrants in Belgium. She later worked on a plain-language newspaper that was accessible to a wide audience; the newspaper helped people at a range of literacy levels (pers. comm., June 5, 2013). Christine Mowat said that plain language helps overcome unequal justice: "We need social change in the law and social change from all levels of government and the professions so that we have less of a class system of language that closes off the public" (pers. comm., June 13, 2013). Mowat also mentioned the wide range of literacy levels in her home country, Canada: one in five residents is foreign-born. Mowat said that plain language can benefit those trying to learn a new language and that plain-language advocates should keep the needs of second-language learners in mind (pers. comm., June 13, 2013).

Four professionals pointed out that ethics is not unique to plain-language work. Carolyn Boccella Bagin, for example, said that she views her ethical responsibilities as no greater and no less than those of any other professional providing a service for a fee (pers. comm., May 30, 2013). Simon Hertnon (pers. comm., May 28, 2013) and Robert Linsky (pers. comm., June 13, 2013) wrote that everyone has obligations to be ethical.

Two specifically said they would not accept work from clients doing work they considered unethical or from clients wanting an unethical service. I suspect most of the others, if not all, would react similarly. As Karen Payton wrote, "Fortunately, I have only been in one situation where I was asked to 'play with words' in order to downplay negative issues. I refused the work" (pers. comm., May 27, 2013).

Identifying Influences on Views of Ethics

> *Prompt A3*: What are the core values and beliefs that affect your views of ethics? Have any specific people, institutions, events, artistic works, or other things strongly influenced your understanding of ethics?

The professionals cited personal experiences most frequently among the influences on their views of ethics. Frances Gordon and Caryn Gootkin, natives of South Africa, cited that country's transition from apartheid to democracy in 1994 (as did Audrey Riffenburgh, an American). Wrote Gordon, "Empowering people through information was key to the new society we were trying to create. So the whole culture in South Africa at that time had a large impact on my understanding on ethics—and how they relate to the business world" (pers. comm., May 21, 2013). Other professionals mentioned that they had learned from situations in which others behaved unethically.

I described the model of BUROC situations in chapter 1. I propose that as a situation reflects more of the traits in the BUROC model (bureaucratic, unfamiliar, rights oriented, and critical), readers in that situation are more likely to benefit from information presented to them in plain language. Audrey Riffenburgh's experience in a BUROC situation affected how she views ethics and plain language:

> My college-educated father was diagnosed with leukemia in 1993 when I was working on my master's. My mother was brilliant but had dyslexia and could not read well. My parents and I were plunged into the medical environment with his diagnosis. In the first 24 hours, we were inundated with papers and forms we were expected to read, understand, decide on, and have my father sign—without help. We were in deep shock and fear and were undoubtedly cognitively-impaired. Being asked to read through pages and pages of dense, medical and legal information before my father could be taken into a badly-needed surgery was excruciating. Giving patients information at that level of difficulty at a time like that is unethical, in my view. This experience was, in large part, the genesis of my decision to work in the medical community to help them understand how to communicate more clearly with patients. (pers. comm., June 25, 2013)

After personal experiences, the professionals most frequently said other people influenced their views on ethics. Four mentioned family members, which is not surprising. In Dragga's (1997) study of 48 working technical communicators, most of them said their family members affected how they understand ethics. Kathryn Catania, for example, mentioned her grandmother, "who always went out of her way to help others. Her motto was to make sure you always do right by the people who need you" (pers. comm., June 28, 2013). Christine Mowat cited her father: "His empathy and compassion for the poor, illiterate, and those who had disabilities affected me" (pers. comm., June 13, 2013). Beyond family members, Caryn Gootkin cited Nelson Mandela (pers. comm., June 11, 2013); Frances Gordon mentioned Paulo Freire, whose work she read in college (pers. comm., May 21, 2013).

Three said that personal values shape how they view ethics. Deborah Bosley said she values fairness (pers. comm., June 17, 2013); Sarah Fox said integrity is important to her (pers. comm., June 15, 2013). Simon Hertnon views ethics through what he describes as universal human needs: survival and betterment (pers. comm., May 28, 2013). He describes his views in detail online (Hertnon 2011). Four others said their religious backgrounds influenced their views of ethics. Said Peter Butt, "For me, the core value is the 'golden rule'—do to others what you would want them to do to you. This has both a theological and a practical basis—and so can be understood from either standpoint" (pers. comm., July 3, 2013).

Investigating Influences of One's Home Country on Views of Ethics

> *Prompt A4*: Does something about your country's unique history, population, or circumstances influence your views about ethics? If so, please explain.

I included this prompt in the questionnaire to see how responses might vary across the geographically diverse group. Although one's home country is not a predictor of one's ethical views, most responded that their home country does influence their views on ethics. Although three from the US reported no influence, two did. Of these two, Annetta Cheek said, "I think of the fact that we are a society that developed from many different peoples and cultures. Honoring that diversity and that heritage is important to me. I think it's something that a lot of Americans have drifted away from, which disturbs me" (pers. comm., June 10, 2013). Canadians Christine Mowat, Michelle Black, and Karen Payton said that their country's diversity, inclusiveness, and desire for fair play affect their ethical views. Payton mentioned that it can be hard to separate the influence of her country from that of her family (pers. comm., May 27, 2013).

South African natives Caryn Gootkin and Frances Gordon were influenced by their country's shift from apartheid to democracy. Gordon said she observed how control of information in the apartheid era affected oppressed people (pers. comm., May 21, 2013). Sandra Fisher-Martins of Portugal said the fascist regime in her country's history perpetuated inequality and exclusion in society. As a result, she said, the Portuguese can be reluctant to confront corruption and stand up for their rights (pers. comm., July 8, 2013). Three New Zealanders said their country affects their views of ethics. Said Rachel McAlpine,

> In New Zealand we are particularly aware of the wrongs done by the colonial government to the first nation to inhabit our country, the Maori. It's good to have the Treaty of Waitangi [signed in 1840 by British colonists and Maori chiefs] in the forefront of our laws and government. We have three official languages: Maori, English and New Zealand Sign Language. (pers. comm., June 15, 2013)

Understanding Plain Language as a Civil Right (or Not)

> *Prompt B1*: Using the scale below [figure 3.1], please choose the number that indicates whether you agree or disagree with the next statement.

"Plain language is a civil right."

I included this prompt because it is a motto of the Center for Plain Language, an influential plain-language group in the US. The statement comes from former US Vice President Al Gore (Dieterich, Bowman, and Pogell 2006), who advocated

TABLE 3.2 Responses to questionnaire prompt B1.

Statement	Number of Responses
I strongly agree	14
I slightly agree	1
I am ambivalent	3
I slightly disagree	0
I strongly disagree	0

for plain language as part of the National Partnership for Reinventing Government during the Clinton administration. I also wanted to see if the statement resonated with professionals outside the US. The responses appear in Table 3.2.

In prompt B2, I asked the professionals to explain their responses. Several themes emerged in the responses of those who strongly agreed with the statement. First, five said that individuals have the right to understand information that affects them. As Karen Payton said,

> For too long the onus has been on the audience to understand the message, not on the sender to communicate clearly in a way that the audience can understand. Whether intentionally or inadvertently, language has been used to create silos of knowledge and power. As citizens and consumers we should be demanding that organizations present information clearly in a way that can be understood. (pers. comm., May 27, 2013)

A second theme, supported by five respondents, is that plain language can address issues of injustice and inequality. Said Rachel McAlpine, "Obscure language is used every day by authorities all over the world as a tool of suppression and deceit" (pers. comm., June 15, 2013). A Write Limited staffer wrote, "We should support disenfranchised people, such as people with low literacy, by being plain and clear" (pers. comm., July 1, 2013).

The third theme, shown in four responses, is that governments have obligations to communicate clearly with citizens. Kathryn Catania said, "The public elects government officials and entrusts them to create sound laws and public programs. When those officials fail to communicate in ways that citizens can easily understand, they fail at meeting the needs of the public that they serve" (pers. comm., June 28, 2013). Annetta Cheek summed up her experiences as follows:

> I was a fed for 25 years, so I'm used to working in the government. I worked a lot in regulatory programs, and I worked a lot with people who were being regulated. It just seems to me that when you are in the business of telling people what to do—and that's the government's business, really—you have an obligation to tell them clearly. (pers. comm., June 10, 2013)

A fourth, related theme in three responses is that plain language helps citizens stay informed. As Simon Hertnon said, "Rights and responsibilities go together and being able to understand each other is at the heart of both. We need to understand our laws, our options, our commitments, our agreements, and our well-earned knowledge" (pers. comm., May 28, 2013).

Several professionals questioned the notion that plain language is a civil right. Caryn Gootkin said, "How people communicate between peers or in closed circles is up to them" (pers. comm., June 11, 2013). Others noted that plain language can run parallel with a person's rights without being a civil right itself. Said Frances Gordon, "To me, the consequences of plain language (fairness, justice, equity) are civil rights. Plain language is just a means to get there" (pers. comm., May 21, 2013). Similarly, Carolyn Boccella Bagin said, "I don't think plain language is the thing that has the right. I think it is the people who have a right to understand information that affects their lives" (pers. comm., May 30, 2013). While she affirmed that plain language is a civil right, Deborah Bosley noted that it falls outside the group of civil rights associated with civil-rights leaders such as Martin Luther King, Jr. (pers. comm., June 17, 2013). William DuBay expressed caution about calling plain language a civil right:

> I don't know if it is a "civil" right. The obligation to provide plain language comes from the obligation of truth itself. The obligation to tell the truth is prior to and much greater than rights created by law. People have a right to plain language only to the extent that they have a right to the truth and that is hard to prove. Can people be said to have a right to something they do not know about, that others say will benefit them? I am very suspicious of trying to limit language to questions of rights and obligations. It is more a question, I believe, of teaching people how to communicate effectively. (pers. comm., May 25, 2013)

In prompt B3, I asked the professionals if their home countries affected their responses to prompt B1. I did not see a clear pattern in their responses, so I cannot assume that their home countries affected their views on whether plain language is a civil right. Overall, eleven said that their home country affected their responses while six said that it did not. Four US respondents said no while two said yes. One New Zealander said no while two said yes. One South African said no while one said yes.

Understanding Plain Language as a Means of Acting Ethically (or Not)

> *Prompt C1*: Using the scale [Figure 3.1], please choose the number that indicates whether you agree or disagree with the next statement.
> "Plain language is a means of acting ethically toward the audience of a message."

I strongly disagree	I slightly disagree	I am ambivalent	I slightly agree	I strongly agree
1	2	3	4	5

FIGURE 3.1 Scale for responses to questionnaire prompts.

TABLE 3.3 Responses to questionnaire prompt C1.

Statement	Number of Responses
I strongly agree	12
I slightly agree	3
I am ambivalent	4
I slightly disagree	0
I strongly disagree	0

I included this prompt to see if the professionals linked plain language with ethical action. The responses appear in table 3.3.

In prompt C2, I asked the professionals to explain their responses. Four said that it is ethical to communicate clearly. Said Audrey Riffenburgh, "Because I value honesty, fairness, and transparency, I believe that providing information with clarity and ease of understanding is a way to act ethically" (pers. comm., June 25, 2013). Four others said that plain language is a means of respecting the audience. Sandra Fisher-Martins said: "Real plain language is not just about short sentences and simple words. It's about having a clear message that addresses the needs of your users and communicating it in the way most likely to meet those needs. It's about being responsive, honest, truthful and fair. It's about respecting people's time, needs, capacities and interests" (pers. comm., July 8, 2013).

Two said plain language helps address linguistic discrimination. Said Caryn Gootkin, "In SA [South Africa], we always need to bear in mind that most of our audience is not an English mother tongue speaker. To respect this diversity, we need to use our language in such a way that does not exclude or discriminate against others" (pers. comm., June 11, 2013). Christine Mowat cited the Canadian Charter of Freedoms: "'Everyone is equal before and under the law, and has the right to the equal protection and equal benefit of the law without discrimination.' I think that's very strong support for acting ethically toward the audience of a message" (pers. comm., June 13, 2013).

Several pointed out that plain language is a tool that people may use both ethically and unethically. Said Martin Cutts, "Plain language can mask a malign purpose just as surely as it can disclose a benign one. It is not the eau-de-nil of the writing world. Like news, it is never neutral" (pers. comm., July 12, 2013). Rachel McAlpine said, "I would love to equate plain language with honesty. Some people do. But sadly, it is easy enough to tell lies in plain language" (pers. comm., June 15, 2013).

Some who expressed ambivalence in response to prompt C1 focused on the practical benefits of plain language over any ethical ones. Kathryn Catania said, "Many people might not even think that deeply about it to ponder right versus wrong. Their motivation may simply be 'I need to get this message out to my reader.' The easiest solution is to write in plain language" (pers. comm., June 28, 2013). William DuBay said, "Plain language is a means of communicating effectively with your audience. They might not have a right to your message or even be interested in it. It is to the speaker's interest to communicate effectively no matter what interest the audience has in the message" (pers. comm., May 25, 2013).

Considering the Strengths and Weaknesses of the BUROC Model

In chapter 1, I proposed the BUROC model as a means of identifying situations in which plain language is likely to benefit the audience. In questionnaire section D, I provided to the professionals the description of the model from chapter 1. I then provided two, two-part prompts:

1. Do you think the BUROC traits (bureaucratic, unfamiliar, rights oriented, and critical) provide a helpful framework for identifying opportunities to reach audiences through plain language? Please explain.
2. Are there any other important aspects of plain-language communication situations that this model should take into account? Do you suggest any changes to the descriptions of the BUROC traits? Please explain.

Evaluating the BUROC Model

Many of the professionals expressed support for the BUROC model. Several called the model helpful. Others said it covers a broad range of situations for using plain language. Said Annetta Cheek, "Given that we can't do everything all at once, I think that model would be a good model for saying, let's start with the documents that fall within the parameters of the model" (pers. comm., June 10, 2013). Audrey Riffenburgh, whose family dealt with her father's sudden diagnosis of leukemia, said the model covers much of the situation her family faced:

> It accurately reflects the situations in which I believe plain language is essential. And it names situations in which I find myself getting angry

when information is not clear. For example, the situation with my father was critical—his survival was at stake. We were in an unfamiliar context and dealing with a huge bureaucracy. It was unthinkable to me that the hospital actually expected people to complete the paperwork under those circumstances. Later, I realized it was impossible for many, many families. At that point, I decided to start a business addressing situations like those, especially for the sake of the families with less education than mine. (pers. comm., June 25, 2013)

Some professionals identified shortcomings in the BUROC model. Several worried that the model might limit use of plain language in everyday situations. (Certainly, I do not want to limit plain language in any way.) Kathryn Catania said, "I don't believe there is a situation when a writer would need to look to a model to identify when an audience needs clear, easy-to-read information" (pers. comm., June 28, 2013). A Write Limited staffer said, "I think that assessing every situation for its right to plain language is unethical. It's time-consuming, judgmental, and risky (we'll get it wrong sometimes). How much better to spend that time writing clearly for every situation; then, thinking of the reader is all you have to do" (pers. comm., July 1, 2013). William DuBay reiterated the difficulty he sees in determining an individual's right to information: "Only if people have paid for information or already have some interest in it, can it be said that they have a right to it" (pers. comm., May 25, 2013). Caryn Gootkin noted that the model does not explicitly mention multilingual or low-literacy audiences. "In these cases, even if the situation is not a BUROC situation, plain language is paramount" (pers. comm., June 11, 2013). Two professionals noted that the model does not explicitly mention the specific audience for whom the message is intended. A Write Limited staffer wrote that the term *bureaucratic* seems to exclude medical and legal issues. Karen Payton said the term *rights oriented* can be difficult to understand.

Considering Changes to the BUROC Model

In response to prompt D2, the professionals identified aspects of the BUROC model that they would change. As with the responses to prompt D1, several worried that the model might limit use of plain language. Simon Hertnon said, "I would be wary of highlighting any set of situations for fear of providing unintended permission not to employ plain language in other situations" (pers. comm., May 28, 2013). Said Deborah Bosley, "I think in general, everything should be written in plain language that has to do with my having to make decisions; any transactional information almost by definition should be written in plain language" (pers. comm., June 17, 2013). A Write Limited staffer also worried about limitations in the model: "Any situation where a person is transacting with an organization or company is an unequal playing field—a situation where the individual is at risk of being steamrolled by the larger body. I suggest that, where dealing with individuals, plain English is essential to keep proceedings fair for the smaller party" (pers. comm., July 1, 2013).

Sandra Fisher-Martins and William DuBay called for a clear focus on readers and users of information rather than the situation. Michelle Black noted that some organizations, especially in the private sector, choose plain language only when they think they risk losing something: time, revenue, customer goodwill, and so on. She worried that a model like BUROC might become a list-box to be checked off and forgotten (pers. comm., June 10, 2013).

Several professionals identified concepts that they think the BUROC model should include. Audrey Riffenburgh and Carolyn Boccella Bagin mentioned that stress affects these situations. Boccella Bagin said that stress can make even familiar situations difficult (pers. comm., May 30, 2013). Caryn Gootkin said the model "should provide for those for whom the lingua franca is not a mother tongue" (pers. comm., June 11, 2013). Karine Nicolay also mentioned multilingual users. She also said that some readers may experience disadvantages such as low literacy, lack of education, or other impairments (pers. comm., June 5, 2013). Sarah Fox wondered if the *critical* component could have more than one meaning: one of importance with weighty consequences and another referring to urgent situations with time constraints. Fox also worried about the *unfamiliar* aspect. She said that turgid language in documents we've used before often leaves us less familiar with a situation than we truly are (pers. comm., June 15, 2013). One Write Limited staffer worried that *bureaucratic* is not broad enough to include medical and legal situations.

I will incorporate the professionals' insights into the BUROC model at this chapter's conclusion.

Considering Whether Plain Language Supports I–You Communication

> *Prompt E1*: In the twentieth century, the German philosopher Martin Buber developed an approach to dialogic communication ethics. In his book *I and Thou*, Buber described two ways that people communicate with each other:

1. In the I–You (or I–Thou) mode, each party respects the other, and the relationship is reciprocal. They speak to each other in dialogue.
2. In the I–It mode, one person speaks down to the other in technical dialogue; there is no true relationship between them.

Chapter 2 discusses how Buber's concepts of I–It relationships, I–You relationships, and the narrow ridge are especially salient for plain-language communication situations. In light of this, I asked the professionals if plain language can provide a means to communicate with someone in the I–You mode rather than the I–It mode. Of those who responded to this question, the vast majority agreed that plain language can support I–You communication. Kathryn Catania explained her affirmative response to the question this way: "Plain language has a lot in common with the I–You mode. It is about respecting the needs of your reader and writing in a way they will best understand. It is also about not sounding pompous or elitist by writing in jargon or gobbledygook" (pers. comm., June 28, 2013).

Three main themes emerged in the explanations of the professionals' responses. The most prominent theme is that plain language is most effective when communicators respect and empathize with their audiences. Sarah Fox connected confusing technical documents to a lack of respect for the audience: "Much technical language shows status and condescension. Partly this is because technical experts like to show how clever they are, and they think the use of jargon and technical terms demonstrates their credibility" (pers. comm., June 15, 2013). Peter Butt said effective plain language "requires the communicator to treat the listener/reader with respect—and as essentially an equal" (pers. comm., July 3, 2013). A second theme is that plain language can address imbalances of power between two parties. Caryn Gootkin mentioned a typical doctor/patient relationship as one with an imbalance of power (pers. comm., June 11, 2013). Sarah Fox said that experts often "set themselves on a pedestal" that precludes dialogue (pers. comm., June 15, 2013). Karen Payton said that jargon-filled, one-way messages from governments and organizations often lead to mistrust among their constituents (pers. comm., May 27, 2013). A third theme is that the plain language of dialogue involves more than words alone. It includes design, organization, tone, accessibility, usability, chunking of information—achieving effective plain language involves more than choosing shorter words and writing shorter sentences whenever possible.

Two professionals said that plainness alone cannot create the reciprocity in the I–You mode between organizations and audiences. Rachel McAlpine said, "I am fond of the I–Thou model for rich and honorable communication. However, plain language is not necessary or sufficient" (pers. comm., June 15, 2013). Frances Gordon said that the surface features of a plain text alone cannot guarantee an I–You dialogue: "It really depends on how you define plain language I suppose. If you have a readability-focused definition, then no. If you have a broader, more discursive definition, then definitely" (pers. comm., May 21, 2013).

Considering Whether Plain Language Supports Narrow-Ridge Dialogue

Prompt E2: In other writings [besides *I and Thou*], Buber noted that deep-seated mistrust often separates two parties. Buber's depiction of the "narrow ridge" frequently appears in discussions of dialogic ethics. The narrow ridge is a place where two parties can meet if they regard each other as *Thou* and not *It*. The narrow ridge is a place from which people in a dialogue genuinely listen to each other and remain open to the other's message. Parties in a narrow-ridge dialogue do not have to compromise or relax their convictions. That said, each party respects the other and demonstrates goodwill. Do you think plain language can provide a way to support a "narrow ridge" relationship between two parties that need to communicate with each other but struggle to do so?

A large majority of those who responded to this prompt said that they thought plain language can support a "narrow ridge" between two parties. Several commented that plain language can create common ground. As Karen Payton responded,

> By the time two parties are struggling to communicate there can already be a built up lack of trust and respect. At this point, both parties need to come to the table as equals and to communicate as equals. This can be hard when there is a clear power imbalance in the relationship (a boss and employee for example). There would be more common ground to be found beyond just the use of plain language. (pers. comm., May 27, 2013)

Said Caryn Gootkin, "If you look for common ground, rather than differences, you can find a way to communicate. If you seek to assert your superiority over another, there is no common ground" (pers. comm., June 11, 2013). Rachel McAlpine pointed out that plain language is a useful tool in counseling, dispute resolution, and diplomatic and industrial negotiations (pers. comm., June 15, 2013). A Write Limited staffer took a different view of plain language creating a meeting place between two parties: "I think that communicating in plain language isn't about a 'narrow ridge,' but about leveling the playing field so that everyone in the conversation can understand each other" (pers. comm., July 1, 2013).

Others noted that plain language alone is not enough to address communication problems between two parties. The advantaged party—whether a physician, an administrator, a legal expert, or someone similar—must actively seek to improve communication and create a "narrow ridge" through plain language. Kathryn Catania wrote that the party with the responsibility to communicate must first acknowledge the communication problem (pers. comm., June 28, 2013). Audrey Riffenburgh, who works in hospitals, has found that many experts look down upon those with limited understanding of their medical issues (pers. comm., June 25, 2013). Those experts are not likely to improve their communication without first respecting all those with whom they work.

Several professionals emphasized the importance of keeping audiences in mind. A Write Limited staffer pointed out that a narrow-ridge conversation might be too complex for some observers but appropriate for the parties involved: "Plain language isn't only used on the narrow ridge. It belongs in the valleys too. It's all about the listener or reader. If the two speakers on the narrow ridge are scientists, the language on the narrow ridge might be complex, but they will understand each other" (pers. comm., July 1, 2013).

Conclusion

The professionals who responded to my questions have provided a base from which further discussions of plain language and ethics can develop. Overall, this group said plain-language communicators have ethical responsibilities to their audiences and their employers. Like Radez (1989) and Shimberg (1989b), they said proper professional behavior is ethical behavior. The professionals said they should show honesty, clarity, and truthfulness in their work. Many see plain-language work as an ethical way to help others. Several professionals reflected the feminist ethical perspectives described in chapter 2 by being aware of differences in power between bureaucratic organizations and their audiences and seeking to address them—by communicating in plain language.

Many of the professionals agreed with the statement that "plain language is a civil right," although not all did. Scholars have not previously examined this statement by Al Gore, which the Center for Plain Language has adopted as a slogan. Among this sample of plain-language practitioners, many agreed with Gore's sentiment, but some found it needs more logical support. Some said that governments and others have obligations to communicate clearly. Many said it is ethical to help others understand information that affects them and to which they have rights. But some noted that it can be difficult to assert and define rights to information. The professionals tended to agree with the idea that plain language is a means of acting ethically toward the audience of a message. At the same time, several pointed out that plain language is a tool with both ethical and unethical uses. A plain-language message is not inherently ethical—or unethical.

I asked the professionals to consider whether plain language might help someone communicate to an audience in Buber's I–Thou mode of reciprocal communication and whether plain language could help create what Buber calls a "narrow ridge" (1965; Arnett 1986, 38) between two parties. On the whole, the professionals supported the idea that plain language can support the kinds of dialogue Martin Buber described in his writings. They did note, however, that dialogue cannot develop unless both parties actively seek it.

The professionals reviewed the description of the BUROC model for identifying opportunities for plain language. Many thought the model could help plain-language communicators in their work. Several offered valuable criticisms of the model. I have used the participants' comments to revise the model and to create an introduction (new text is in italics):

> *The BUROC model provides one way to identify situations in which plain language can benefit readers. BUROC situations are important because of the*

challenges they present to people who need to acquire information and then act on it. People going through BUROC situations may feel stress or anxiety; they may be dealing with uncomfortable situations; they may have to use languages other than their native tongues; they might face pressure to act quickly. Plain language can help people facing BUROC situations feel more at ease, understand more about their situations, and make decisions more confidently.

- B is for bureaucratic. These situations involve some kind of bureaucracy; layers of policies, procedures, *and approvals affect individuals' access to what they need or want. Those with expertise and authority in these situations, such as physicians, lawyers, legal experts, or managers, enjoy advantages unavailable to those who need their services.* The bureaucracy's public façade often keeps outsiders distant and limits their access to information. *These situations are complex, they may require a lot of time to resolve, and they may occur over several episodes.* Some bureaucratic processes include making insurance claims *and obtaining treatment for a serious medical condition.*
- U is for unfamiliar. People encounter these situations rarely or infrequently. They require people to use jargon, policies, and even facilities that are not immediately at their command or recollection. *These situations may be unfamiliar because they involve vocabulary that is confusingly complex. They may be unfamiliar because they require people to use a language not native to them.* An ill patient considering enrolling in a clinical trial likely faces unfamiliar terms and concepts. *An immigrant reading a mobile-phone contract in his or her second language may experience unfamiliarity.*
- R and O are for rights oriented. These situations are especially important because they affect individuals' choices to act within their rights—rights as citizens, as patients, as consumers, as humans. *Election officials instructing citizens on how to operate voting machines correctly face a rights-oriented situation.*
- C is for critical. These situations are weighty; they are important; *people should not regard them* lightly; they can have significant consequences for people facing them. These situations often arise without warning, and they may require urgent decisions *or actions. The concurrent stress or fatigue in these situations can affect a person's judgment, cognition, and performance.* For example, a policy document for an organ transplantation network addresses *critical* matters of life and death for individuals needing new organs; administrators *must implement* such policies quickly.

As mentioned in chapter 1, the first edited collection on ethics in technical communication omits plain language (Brockmann and Rook 1989). Coeditor

Brockmann writes that "plain language, although a readability concern, is not necessarily an ethical concern. Identification of plain language with ethical language mistakes the outward signs of ethics, plain language, for true ethical actions" (1989b, v). The professionals in this study agreed that plainness does not necessarily make a message ethical. But they tended to disagree that plain language is not an ethical concern. On the contrary, several professionals agreed that ethics is a part of many aspects of plain-language use. Many of them see plain-language work as a means of ethical action.

Questions and Exercises

1. Who has influenced your views of ethics? Identify at least three people; name at least one person outside your family. Write a paragraph of 50 to 75 words about each person you have named that summarizes that person's impact on your ethical views.
2. Refer to Prompt B1 above. Do you believe that plain language is a civil right? Write a paragraph of 50 to 100 words exploring your views. Take into account arguments for both sides of the issue.
3. Think of a BUROC situation that you have faced. Which part of the BUROC model was most salient to you in that situation? How did the situation get resolved, if it did? Describe your experience in the situation in about 75 to 100 words. Write another 75 words analyzing an ideal solution to that problem. Finally, write 100 words about what it would take to ensure that everyone in the situation you faced received appropriate support.

4

PROFILE–HEALTHWISE, INC.

Healthwise, Inc., is a nonprofit company in Boise, Idaho, that employs writers, editors, and physicians to create health and medical information, technology, and services for consumers in the US, Canada, and around the world. Since 1975, Healthwise has grown from a handful of employees to an organization of more than 250 employees. In April 2014, Healthwise merged with the Informed Medical Decisions Foundation, a Boston nonprofit organization that promotes shared decision making between patients and their care providers by developing instructional materials, supporting research, and advocating for policy changes. Healthwise and the Informed Medical Decisions Foundation now share a board of directors, but they retain their distinct names, locations, and expertise (Healthwise, Inc. 2014).

I studied Healthwise because the organization has demonstrated its commitment to plain language through training its employees, through its culture, and through awards that it has received for its work. Healthwise is also an interesting case because of its size: it is larger than a typical start-up company, consultancy, or nonprofit organization yet smaller than a large firm such as a pharmaceutical company or medical-device manufacturer. I interviewed several people at Healthwise, including vice presidents and the CEO. I also observed meetings of editors and writers.

This profile reveals Healthwise's ongoing concern with pursuing and promoting ethical behavior as a health- and medical-information provider. It shows how Healthwise takes a dialogic approach to communication as it seeks to help people make good health and wellness decisions. I describe how Healthwise's content addresses BUROC situations, who creates plain-language content at Healthwise, how Healthwise creates its content, and how organizational culture and ethics affect Healthwise's work. I close with lessons that plain-language professionals may take away from Healthwise.

Background of Healthwise

From the beginning, Healthwise has had a consistent mission: to help people make better health decisions. Don Kemper launched Healthwise in 1975. Healthwise grew out of Kemper's work with another nonprofit called Health Systems Incorporated (HSI), then a project of the National Centers for Health Services Research. Kemper says that the simplicity of the Healthwise mission has allowed the company to succeed over several decades: "It is a true gift to have that clear direction that allowed us to be successful over this long period of time even while the world around us changes dramatically. From a position of innovation, we're able to look at everything changing in the world through a lens of 'How does it help people make better health decisions?'" (pers. comm., April 3, 2013).

Healthwise licenses its content to a variety of organizations, including hospitals, health-insurance companies, disease-management companies, government agencies, and web portals. Rather than selling products to consumers in what many call the business-to-consumer (B2C) model, Healthwise sells its products to other businesses in the business-to-business (B2B) model. These businesses then share the content with their customers. Miriam Beecham, vice president for product management, described the market for Healthwise products as "B2B2C" (pers. comm., April 1, 2013). Although Healthwise's business clients are intermediaries who distribute the content, Healthwise develops all content—including videos and other interactive media—with the end consumer in mind.

Kemper studied engineering sciences as an undergraduate in the 1960s, and then he served two years in the US Public Health Service. During his service, he attended a lecture by Dr. Vern Wilson, a physician and then an assistant secretary of the Department of Health, Education, and Welfare. Wilson said that the greatest untapped resource in health care is the patient. Kemper took Wilson's statement as a reason to empower patients instead of people in the medical profession. Wilson's statement inspired Kemper to seek ways to help patients identify and use their own resources in their own care. Around this time, Kemper had an infant child, and he had received a copy of Dr. Benjamin Spock's bestselling guide on raising a child, *Baby and Child Care*. Kemper thought that someone should produce a similar guide about self-care for an entire family: what to do for colds and flu, headaches, and backaches; how to prepare for an effective visit to the doctor; and so on. Kemper said he was unable to convince the federal government to write such a book, but he held on to the idea. Kemper earned a master's degree in public health before joining HSI in Boise, Idaho (pers. comm., April 3, 2013).

Throughout his career, Kemper has sought to help people with their health decisions in a variety of ways. After moving to Boise, Kemper set up interviews with people in southwest Idaho and eastern Oregon. Interviewees confirmed that they would like to know more about self-care. Kemper said, "They basically told us the same thing: help us decide what we can do at home, tell us how to prepare for visits to the doctor and get our money's worth, and help us understand what role we have in our health-care decisions" (pers. comm., April 3, 2013). At HSI, he helped develop a class called The Patient's Role in Health Care, which the company presented in the adult-education program of the Boise public school system. HSI also created an early attempt at interactive media, developing "Common Sense, Common Health" with a Boise television station. Each of 10 episodes covered an aspect of primary care and featured information from an appropriate health-care provider. HSI formed viewing groups in the community—in church groups, parent-teacher organizations, and the like—to watch each show. Viewers then called into a radio program after each television broadcast to ask the featured provider questions.

At Healthwise, Kemper was able to create the self-care manual that he could not convince the government to produce: *The Healthwise Handbook*. Insurance companies, medical practices, and other large organizations have widely distributed the book, now in its eighteenth edition, to consumers. Over time, Healthwise has used many different media to deliver health information to the audience. The Healthwise Knowledgebase is a large repository of health information that consumers can use through a web portal, such as WebMD or hospital and health-plan websites. Healthwise creates instructions that care providers can give to patients before and after specific treatments or encounters with the health-care system. Videos provide information to supplement the written word. Online coaching tools and interactive modules give patients information appropriate for their specific circumstances.

From the beginning, Kemper insisted on creating content that reflects current evidence-based medicine (pers. comm., April 3, 2013). The content team, described in more detail below, includes writers, editors, and managers. Many content team members have training in the sciences; several have worked as nurses and in other clinical roles. Healthwise has been a leader in the practice of employing physicians and other qualified care providers to review content for medical accuracy and to ensure it reflects medical practice. Dr. Martin Gabica, chief medical officer and a board-certified family physician, leads the medical-review team. Medical reviewers include physicians and a wide range of medical specialists. Healthwise operates its own medical library and actively monitors news and research in evidence-based medicine.

In addition to medical accuracy, Healthwise has always worked to create information that consumers can easily read and understand. Karen Baker, the senior vice president for consumer experience, oversees the content and user-experience teams. Baker said that Healthwise was writing in patient-friendly language before the term "plain language" appeared widely in discussions of health literacy (pers. comm., January 31, 2013). Baker and Kemper each said that it has always been important for Healthwise to use common, accessible language to help consumers be more actively involved in their own health care. Kemper said it would be "unconscionable" for Healthwise to give consumers information that left them confused: "It's really all about informed decision making. The goal isn't plain language; the goal is informed decision-making or informed action" (pers. comm., April 3, 2013).

After joining the company in 2001, Baker suggested that Healthwise content would be more effective if "we wrote *to* the patient, using second person, rather than *about* the patient, using third person" (pers. comm., January 31, 2013). Baker's suggestion about writing directly to patients cultivated a budding interest in plain language among Healthwise employees. She brought in an expert to provide formal instruction in plain-language writing and gradually added proven plain-language practices to the content team's processes. In 2006, Baker led the launch of a user-experience team that carries out one of the central tenets of plain language: gathering feedback to ensure that the information "actually *works* for the people who use it. It starts with the audience. It's not only the words we choose; it's the design, the format, and the user research and testing that tell us we are hitting the target—or not" (pers. comm., January 31, 2013). External awards show that Healthwise is committed to plain language. The Center for Plain Language has given annual ClearMark awards since 2010 for outstanding examples of plain language. Healthwise has won four ClearMark awards, including the inaugural Grand ClearMark in 2010; each award recognizes Healthwise's work in a different medium.

Healthwise operates a proprietary system that tracks the number of times consumers use or "touch" the company's content. An electronic display in the main lobby shows that the total of touches exceeds one billion and continues to grow. The display reminds Healthwise employees that consumers trust the company's content and use it to make informed decisions.

When consumers need to make decisions about their health, they face situations that are bureaucratic, unfamiliar, rights oriented, and critical (BUROC) in one way or another. Tad Arnt, vice president for client services, said that the bureaucracy of health care is often set up for individual episodes of care, such as a patient's visit to a particular doctor. But when a patient needs care from a series of providers and facilities, the patient (and the patient's family) must navigate the bureaucracy largely independently (pers. comm., March 19, 2013). Baker, the senior vice president over the content and user-experience teams, and CEO Kemper each said that consumers' unfamiliarity with health and medicine sustains a power differential. Professionals in health care and medicine have more knowledge and more power than patients do. Health situations that are urgent and critical can affect individuals' mental health as well as their physical health. Christy Calhoun, the vice president of content, said that the uncertainty and anxiety associated with a serious condition can leave an individual overwhelmed and unable to act. Calhoun said Healthwise tries to understand the patient's experiences and present content that helps readers address their problems in real contexts (pers. comm., March 26, 2013).

Patients have rights to make decisions about the care they receive, but sometimes they have trouble exercising them. Patient preferences and values matter. In some cases, patients may need to choose between taking medication over a long term, having surgery, and significantly changing their diet and exercise routines. In other cases, patients might feel pressured by a doctor to undergo surgery when other treatment options are available. In addition to providing detailed information on thousands of health and medical topics, Healthwise produces many decision guides to help patients think about their preferences about care and to understand and exercise their rights to choose between options for care.

Personnel Who Create Plain-Language Content at Healthwise

Because information is the core product that Healthwise produces, a significant number of employees support content-development efforts. Although the content team supports Healthwise's proprietary content-development processes, I describe the members of the team in general terms to provide a sense of the team's work:

- *Senior vice president of consumer experience.* This senior vice president oversees the work of the content team and the user-experience (UX) team.
- *Vice president of content.* This vice president leads the work of the content team. The scope of the content team's work encompasses writing new and enhanced content across all Healthwise products, keeping content medically accurate and consistent, ensuring that the team develops content according to Healthwise style guidelines and standards, and localizing content for Canadian audiences.
- *Medical-content specialists.* These individuals help make decisions about content development in particular areas, such as cardiovascular health, pediatric care, or mental health. They work with in-house medical reviewers to decide when to create new content, when to revise existing content in light of changes in current medical evidence, and what is important for a patient to know.
- *Writers.* They translate current medical research into understandable and usable content for a variety of media, including video, print, websites, and interactive content.
- *Associate editors.* They proofread and edit Healthwise content in several media. Associate editors edit for correctness and consistency in style and usage, plain language and readability, and appropriate formatting. Associate editors also provide training within the company to help others create correct and consistent content.
- *Editorial managers.* These individuals ensure content is appropriate for both the audience and the product that delivers it. They provide substantive editing and may make recommendations on the structure, organization, and tone of Healthwise content. Editorial managers tend to have substantial experience creating content at Healthwise; many also have clinical experience as health-care professionals.
- *Media developers.* These programming and design specialists put content into the appropriate medium.

The content team works with the medical-review team to ensure that content is accurate and appropriate for a consumer audience. In-house medical directors are reviewers who consult with writers and editors frequently; often they are board-certified family physicians. Specialist reviewers are part of an extended, external network of providers who practice in particular areas of health and medicine. Specialist reviewers complete their work remotely. The UX team conducts usability testing so that content and products reflect user needs and preferences. The UX team also provides expertise in areas that benefit content and products, including medical illustration, video production, animation, visual design, interaction design, metadata, and information architecture. The content team works on hundreds of projects at once. Healthwise assigns a project manager to support each content project. Project managers help team members solve problems and meet deadlines.

Practices and Processes for Creating Plain-Language Content at Healthwise

Healthwise uses many practices to help the content team produce effective plain-language materials. These activities include training, planning, writing with personas, working in teams, following strong editorial practices, user testing, and understanding readability holistically.

Promoting Plain Language through Training, Practice, and Leadership

Healthwise promotes plain language through training, practice, and leadership from key people. Every new content team member receives an introduction to plain-language principles from Michele Cronen, the lead associate editor. Cronen said she provides statistics such as those from the National Assessment of Adult Literacy to show why plain language benefits a broad audience and how it supports the Healthwise mission. The team that Cronen leads also provides training seminars to content team members and others about plain language and other aspects of writing (pers. comm., March 1, 2013).

Over the past few years, Healthwise has developed an approach to content development called the three Ps: plain, personal, and possible. Each one of the three Ps affects the other two. In terms of plainness, the content team works to write consistently in plain language. Personal content is content that resonates strongly with a reader; Healthwise wants to write for readers without talking down to them. The third P, possible, means that Healthwise hopes readers will act on what they read. In many situations, consumers have health or medical issues that require them to make significant changes: taking medication regularly, exercising more frequently, changing what they eat and drink, and so on. Content team members understand that these changes can be difficult. Cronen said that the best writing will empathize, motivate, and inform. She explained that Healthwise content developers "strive for the tone of a knowledgeable and supportive friend, one that helps people feel like they are getting information from a trusted, expert source that is approachable, and not too authoritative" (pers. comm., March 1, 2013). At the same time, said Christy Calhoun, the vice president managing the content team, writers also try to avoid a tone that is overly positive and glosses over the seriousness of a patient's situation. Healthwise employs a behavioral psychologist to help the team develop content that readers think is possible for them to act on (pers. comm., March 26, 2013).

Strong, visible leadership also promotes plain language at Healthwise. CEO Kemper said that for Healthwise to fulfill its mission of helping people make better health decisions for themselves, the company must create content in language that consumers can easily understand. Thus, support for plain language at Healthwise starts at the top of the organization (pers. comm., April 3, 2013). Several people I talked to identified Karen Baker, the senior vice president of consumer experience, as a key supporter or champion of plain language. Kemper said that Baker has been especially effective at sharing the Healthwise perspective on plain language outside the company. For example, Baker speaks at conferences and judges entries for plain-language award competitions. Kemper was quick to add that Healthwise has many champions of plain language on Baker's team (pers. comm., April 3, 2013); lead associate editor Cronen is probably foremost among them. After Baker brought in the plain-language trainer years ago and sought people to support plain language at Healthwise, Cronen quickly volunteered. After working on plain language at Healthwise for several years, Cronen has become "the plain-language person" to her colleagues.

Cronen leads a monthly norming meeting in which members of the content team and the medical-review team can discuss recent experiences writing with the three Ps. Team members typically review content they have completed, but sometimes they review content that is still in development. These discussions let team members share ideas about creating plain-language content effectively, and they reinforce plain language as part of the Healthwise culture. These norming meetings tend to focus on identifying lessons learned from content that the team has produced and applying those lessons in the future. Cronen records minutes from each meeting and distributes them to writers, editors, reviewers, and product managers (Michele Cronen, pers. comm., March 19, 2013).

This example shows how team members at Healthwise applied the three Ps to a message, dropping the grade-level score from sixteenth in the original to sixth in the revision:

> Original: Stress can negatively affect your heart in many ways, but you can lower your stress level through talking about your problems and your feelings, exercising, and doing deep breathing, meditation, or yoga.

Revised: Stress can hurt your heart. Keep stress low by talking about your problems and feelings, rather than keeping your feelings hidden. Try exercise, deep breathing, meditation, or yoga. (Christy Calhoun, pers. comm., March 26, 2013)

The revised text differs from the original in several ways. The original text is a single compound sentence while the new text has three sentences. These sentences are much shorter and more conversational than the original. The initial statement in the original, "Stress can negatively affect your heart in many ways," is now more direct and personal: "Stress can hurt your heart." The original list of suggestions for lowering stress has one very long item followed by several shorter ones. The second sentence of the revised text better explains that hidden feelings can raise a person's stress. The third revised sentence provides a short list of tactics that a person can investigate and try.

Planning

Planning is an important aspect of the content-development processes at Healthwise. Every piece of content that Healthwise creates begins with a content plan. Content specialists and writers create plans together. A plan describes the purpose for the content, the intended tone of the content, the target audience, and the objectives that readers should be able to accomplish after reading the content. The plan will also state whether the team should create the content with a target persona in mind. These plans are available to all team members during the content-creation process, and team members use them often. Planning also involves scheduling individual projects and setting milestones to gauge progress.

Writing with Personas

Within the past decade, Healthwise created a UX team to conduct research to better understand how consumers use Healthwise content. By reading published research and using the UX team's findings, Healthwise has created personas to help in developing content and products. Personas describe the characteristics of a target user of an application or a piece of content. A persona combines traits that real people exhibit into a description of a fictional person. Calhoun, vice president of content, described how personas influence the content-development process:

One of the personas we've developed recently is for coronary artery disease. Based on interviews we conducted with people who had a recent heart attack, we gained terrific insight into the emotional struggles of that health journey. In that body of research, we've found that many people are surprised to learn that they were at risk for a heart attack. Or if they've had a heart attack and that was the first health crisis that awakened them to the fact that they had a health issue, we learned that they were completely taken aback by that. Some words that patients used to describe that situation were, "I thought I was invincible; I'm so afraid now; why didn't anyone tell me I was at risk?" We try to understand what words patients are using to describe their conditions so that we can use similar words to frame the content and address the fears they might be experiencing and help them overcome that anxiety. So, this development of personas has helped us to tailor the content and the experience to make content feel more personal to their unique situations and to help make lifestyle changes seem possible. (pers. comm., March 26, 2013)

These personas reinforce the fact that Healthwise audience members must confront not only the facts about their medical situations but also the attendant feelings and emotions. The personas represent people facing specific BUROC situations.

Working in Teams

Each content project at Healthwise involves several people. All people involved, from the content team to members of the medical-review team, have the opportunity to shape the content in plain language. Team members collaborate frequently, and a flat organizational structure encourages everyone to contribute. Even though some team members have medical degrees or doctoral degrees, team members use first names to address each other.

Clinicians, whether they are medical reviewers or members of the content team, bring important experiences that help them know when a piece of content might not be plain enough for patients. The generalist physicians on the medical-review team have most likely had to explain a given concept to patients in practice. Their experiences give them a sense of when a definition of a complex term or procedure might confuse a patient. Gabica, the chief medical officer, said this about how he talks to writers:

Often I will find myself saying to a writer, "This is how I used to explain this to people." Because in dealing with patients, I had to explain things to them and then check that they understood it, and redo that if they didn't. And that's gotten me over the years to sort of some standard language that I've used to explain things to help them understand it better. And that turns out to be pretty helpful sometimes here. (pers. comm., March 1, 2013)

Healthwise's medical reviewers help writers understand the dynamics of conversations between physicians and patients. Through experienced clinicians like Gabica, writers can begin to understand the subtleties of physician–patient dialogue.

Gabica also pointed out that medical-review team members are not the only personnel with clinical experience. Several members of the content team have worked in nursing, physical therapy, and other related areas, and they use their experiences to shape the content they produce.

Iterative processes help the content team to produce content in plain language. Gabica acknowledged the challenge of conveying complex ideas in plain language and said team members work iteratively until a piece of content is effective (pers. comm., March 1, 2013). Small nuances in word choice have important effects, and some phrases or longer passages may go through several drafts. Cronen, the lead associate editor, said the team works to refine a piece of content until it is plain, accurate, and helpful for the end user. Sometimes she and a writer will delete a piece of content and start over if it is not conveying information effectively (pers. comm., March 1, 2013).

Some amount of tension between plainness and precision is inherent in the review process. Specialist physician reviewers may at times suggest changes that make the language more complex. This tension is understandable; specialists routinely use complex vocabulary in their work. Healthwise includes specialists in the review process because their perspectives are important and because specialist review is a best practice for developing health content. If reviewers disagree about using a particular term or phrase, team members will choose the solution they think best serves the end users of the content. But that does not mean that Healthwise will avoid complex words or complicated concepts. Calhoun said that Healthwise teaches complex terms to empower patients so that they can feel more confident about managing their health and not feel as overwhelmed when they are in the doctor's office. For example, Calhoun said that a patient with diabetes needs to know about neuropathy, nerve damage caused by high blood sugar. A patient with heart problems may need to understand arrhythmia, an abnormal heart rhythm (pers. comm., March 26, 2013). Baker, senior vice president for consumer experience, said that Healthwise might sacrifice precision to make a piece of content more readable and usable, but Healthwise will not compromise the medical accuracy of its content (pers. comm., January 31, 2013). As chapter 2 mentions, the concern for accuracy and desire to avoid misleading readers are central to professional ethics in technical communication. Healthwise acts ethically by seeking to provide accurate health information that consumers can understand and act on.

Following Strong Editorial Practices

The content team follows the company's editorial practices, and the associate editors take a lead role in supporting those practices. In addition to performing typical copyediting duties of proofreading, editing to ensure correctness, and enforcing formatting standards, Healthwise associate editors ensure that content reflects principles of plain language and health literacy. Healthwise's processes ensure that associate editors have authority to modify content so that it is appropriately plain.

The associate editors work actively with content team members to promote shared understandings of plain-language principles. Several people I interviewed, including three vice presidents outside the content team, mentioned the company's plain-language glossary. This glossary, maintained by the associate editors, collects medical and health terms that people throughout the content team have translated into plain language. The glossary saves time for writers and also helps promote consistency across the many forms of content that Healthwise produces. Cronen, the lead copyeditor, regularly sends out a short email newsletter with reminders about plain-language resources. A typical newsletter will identify specific entries in the plain-language glossary and will include links to news stories about plain language. The efforts of the associate editors link the content team's editorial practices with the company's culture.

Understanding Readability Holistically

Healthwise believes that the reader is the best judge of whether a piece of content is readable. Some Healthwise clients specify that they want content that meets a particular grade-level standard, so content team members work with authoring software that can calculate a grade-level readability score. Grade-level scores tend to go higher (that is, to indicate more difficulty in reading) as writers use longer words, a relationship that could discourage writers from using medical terms. Miriam Beecham, vice president of product development, said Healthwise puts the needs of readers over specific readability scores. "Our feeling is that we could bring the grade level down even further by removing the name of the condition or removing the name of whatever it is that [readers] have from the article, but then that doesn't give the patient or the consumer a 'crosswalk' between what they're hearing the physician say and what they need to understand" (pers. comm., April 1, 2013). Tad Arnt, vice president for client services, said Healthwise would "do a significant disservice to consumers" if the company left out important medical terminology for the sake of reaching a particular grade-level score (pers. comm., March 19, 2013).

Calhoun said that some clients agree with Healthwise that readability involves more than grade-level scores. These clients, including some state Medicaid agencies, also look at the following aspects of content:

- The presentation and layout of the content, including use of white space, bulleted lists, and boldface text
- The use of images and multimedia to convey complex topics or to provide engagement
- The extent to which the layout helps readers easily scan and read the information

Calhoun said, "They don't just look at the reading level. They look at all of these other important factors that help people to understand information at the most granular level" (pers. comm., March 26, 2013). Content team members take all of these factors into account and approach readability holistically. Synthesizing many decades of research and his own decades of experience, Kimble (2012) provides holistic guidelines for creating readable documents in plain language. The guidelines cover not only words but also sentences, organization in paragraphs and larger sections, and page design (5–10). Kimble's guidelines overlap substantially with Healthwise's practices.

Translation is another aspect of readability for Healthwise content. Healthwise creates content for the US market in American English and in Spanish. Content for Canada is in Canadian English and in Canadian French. Calhoun said Healthwise translates a set of its content for Medicaid clients into the 10 languages that immigrants to the US most commonly speak, including Chinese, Vietnamese, Russian, Somali, and Farsi. Calhoun said these translations help people without strong English skills (pers. comm., March 26, 2013). Healthwise's investments in translations show that the company continually strives to meet the audience's needs no matter what languages readers use.

Organizational Culture and Plain Language at Healthwise

Healthwise has a strong corporate culture with a focus on ethics. CEO Kemper told me that anyone I stopped in the hallway would most likely be able to name the company's three core values (pers. comm., April 3, 2013). I heard the principles so often in my interviews that I had to agree with him. These values are respect, teamwork, and "do the right thing." Kemper said that after Healthwise crossed the threshold of employing 50 people in 1995, he asked employees about traits they wanted the company to reflect. Kemper read through the responses and identified the three core values (pers. comm., April 3, 2013).

Teamwork is an important part of plain-language culture and practices at Healthwise. Teams, rather than individuals, produce each piece of content. These teams tend to come together according to subject matter or specific markets. For example, the same team members tend to write, review, and copyedit all new content about diabetes. Another team reviews and updates existing content on a regular schedule. Content localized for the Canadian market usually comes from an assigned group of writers and the same copyeditor each time. This consistency in team membership allows employees to build fluency with the subject matter and to build effective working relationships. At the same time, employees work on several project teams at once, so they become familiar with the work of a larger group of people.

Respect is an important part of working on teams and of doing the right thing in any situation. Because Healthwise employees collaborate so frequently, mutual respect is especially important. Healthwise conducts an anonymous satisfaction survey of employees twice each year, and several questions relate to the respect that employees experience within and among their teams. Healthwise employees extend respect to their audience as well: They use the three *P*s of plain, personal, and possible to show respect to the audience and to be aware of the audience's concerns.

"Doing the right thing" is a concept that several people mentioned to me. Kemper said that Healthwise employees share the responsibility of acting rightly and ethically:

> We tell our employees in orientation and overall, if you make a decision, if you take action because you think it's the right thing to do—whether it's for a client, or for a supplier, the mail person, or for a coworker, or for yourself—if you do it because you think it's the right thing, we will back you up. You do not have to stop and ask. We may then tell you next time, there's a different way to do this. But we believe there is far more benefit when you are taking on the responsibility of deciding what's right rather than going through a chain of decision makers. (pers. comm., April 3, 2013)

Healthwise behaves ethically and humanely toward its employees by empowering them and taking a dialogic approach toward ethical decisions. This stance contrasts sharply with the three corporations whose codes of conduct Dragga (2011) examined. Although the Ethisphere Institute considered the three corporations ethical, Dragga showed that they frequently took a monologic, coercive approach toward shaping employee behavior.

Healthwise holds monthly all-employee meetings. Although I did not attend any of these private meetings, I learned that they reinforce the company's culture, including its support for plain language. Several sources mentioned that if a speaker uses too much jargon in one of these meetings, the audience will gently chide the speaker to use plain language instead. The meetings also include stories or vignettes about how clients or consumers use the company's plain-language content. Arnt, vice president for client services, recalled a story told by a Healthwise colleague in software engineering (pers. comm., March 19, 2013). This colleague was spending time with a husband and wife in Cascade, a small resort town about 90 minutes north of Boise. The woman felt ill, but she did not seek care because she thought she only had heartburn. Arnt's colleague used his phone to open an online Healthwise module about cardiac symptoms. According to the module, the woman's symptoms matched those of a heart attack. They immediately took her to a local clinic, and medical personnel found that she did have a heart attack. The clinic treated her successfully, and she recovered—after the information from Healthwise prompted her to seek care. Stories like this motivate Healthwise employees and remind them that the company's mission is important. Both external and internal audiences respond to a company's brand. Arnt went on to say that a company's brand, which figures prominently in its advertising and marketing, is the company's identity. Arnt said that plain language is part of the Healthwise brand, inside and outside the company (pers. comm., March 19, 2013).

While several leaders in the company openly and frequently support plain language at Healthwise, the associate editors actively support much of the day-to-day plain-language activity. Associate editors provide periodic training to writers, and they manage the norming meetings for the three *P*s. The associate editors also have a tradition of celebrating National Grammar Day every spring. They set up activities in the company's main lobby, where most employees enter and exit the building. In 2013, the associate editors focused on using verbs in the active voice, and their activities coincided with a theme of physical activity. Records show that more than one-third of Healthwise employees usually participate in National Grammar Day. As International Plain Language Day becomes more prominent in October of each year, the editing team will help Healthwise celebrate it in similar ways (Michele Cronen, pers. comm., March 1, 2013). In October 2014, Cronen and Baker marked International Plain Language day by offering a public seminar on plain language and health literacy at Boise State University. Events on National Grammar Day and International Plain Language Day range from the fun and lighthearted to the serious and informational. By keeping these events on the company's calendar, Healthwise's associate editors help ensure that plain language is part of the company culture.

Linking Ethics and Plain Language at Healthwise

Kemper and Healthwise have long been leaders in ethical conduct for providers of online health information. Kemper said a 1999 scandal involving the website DrKoop.com, whose founders included former Surgeon General C. Everett Koop, prompted people in the industry to develop ethical standards (pers. comm., April 3, 2013). A story in the *New York Times* revealed that Koop's website blurred the line between advertising and editorial content in the way it presented organizations that paid to appear on the site (Noble 1999). Kemper said he, Koop, and others joined to create a group called Health Internet Ethics, or Hi-Ethics. The group selected Kemper as its chair (pers. comm., April 3, 2013). Hi-Ethics developed 14 principles to help organizations create quality, effective health information while behaving ethically. These principles included developing privacy policies to protect site users, disclosing financial interests, identifying writers and reviewers of information, and allowing consumers to provide feedback on the information they have read (Mack and Wittel 2001). After about two years, Hi-Ethics dissolved and turned over its work to an accrediting body called URAC. Once known as the Utilization Review Accreditation Commission, URAC is an independent organization that provides accreditation for health websites and many health-services organizations. Calhoun says that Healthwise maintains URAC accreditation as a benefit for its clients (pers. comm., March 26, 2013).

For Healthwise, plain language is part of ethical behavior. Since the mission of Healthwise is to help people make better health decisions, plain language, Beecham says, helps Healthwise to do well in its mission and to do good for consumers:

> From that perspective, we have an obligation to use plain language in order to meet the mission as best we can with the resources that we have. If we weren't using plain language, we would be reaching a much smaller percentage of the audience out there, and it would really be a waste, an inefficient use of the resources and the great minds that we have here to do that. (pers. comm., April 1, 2013)

Beecham's comment shows not only a concern for the ethical concept of utility, seeking the best results for the most people, but also a sense of care for consumers. This sense of care connects directly with the company's mission to help people make better health decisions.

Several people with whom I spoke at Healthwise said that doing the right thing is an effective way to describe ethical behavior. Beecham said ethics is doing the right thing by the people that you're serving or you're helping (pers. comm., April 1, 2013). Calhoun said that members of the content team have an ethical obligation to consumers and that they support the company's mission by focusing on consumers' needs for information (pers. comm., March 26, 2013). Gabica, the chief medical officer, pointed out that a physician's ethical obligations toward a patient do surpass a writer's or editor's ethical obligations toward a reader. Nevertheless, he did say that Healthwise personnel have an ethical obligation to give consumers information that is evidence based and that they can understand (pers. comm., March 1, 2013).

Because Healthwise is a nonprofit company, it can put its mission ahead of its profits. Kemper said that nonprofit status allows Healthwise to focus on long-term objectives rather than short-term financial goals. The product-development team continually looks ahead and tries to develop products and services that will lead the industry. As an independent company, Healthwise strives to provide unbiased information. It does not have relationships with pharmaceutical companies, hospitals, or other groups who might want to influence its content. Kemper said that no one can purchase Healthwise because the company is not for sale (pers. comm., April 3, 2013). The company succeeds when it helps people make better health decisions, not when it reaches sales quotas. This arrangement helps Healthwise put full energy and effort into helping people. Healthwise also acts ethically by avoiding relationships that create conflicts of interest.

Supporting Dialogue with the Audience

From the beginning, Healthwise has supported a dialogic view of communication ethics. Healthwise has treated health consumers as "Yous" and not "Its" by using accessible language and by providing information on health and medical topics that consumers are likely to encounter.

Healthwise takes several steps to gather input and feedback from those who use its products. One of the primary means of supporting dialogue is through the work of the UX team. The UX team conducts research on Healthwise content with members of the target audience. Some research is generative, to identify content or products that will meet audience members' needs. Summative research tests the effectiveness of finished products. Becky Reed, senior director of user experience, described user testing at Healthwise:

> We do both upfront research with patients before we design—to help shape the vision of what we do—and then user-test our content and software as we develop it. We compare what patients told us about their journey with our content and see how it looks—will this work engage a patient who has a lot on their mind, will it make them more confident to do what they might need to do, will they do something differently or keep going as a result of their experience with Healthwise? This is about providing the most relevant health information to the patient who needs it and making hard issues seem possible—and it's impossible to do that without listening to patients first. (pers. comm., July 19, 2013)

Healthwise respects its audience and fosters a type of dialogue by gathering feedback from them through usability testing. This feedback also helps reduce the power differential between consumers and medical experts. Clients who provide Healthwise content to consumers also give feedback about their experiences using the content. Clients often provide helpful suggestions about new content to develop, and they also help identify ways to improve existing content.

To support dialogue with the target audience, Healthwise involves writers, editors, and reviewers who know consumers well. The in-house medical directors are physicians with practical experience in helping patients understand the issues that Healthwise content addresses. Editorial managers often have experience as nurses, physical therapists, or other types of clinicians. These experienced clinicians stand in for target audience members when it is not practical or possible to consult with consumers themselves.

Conclusion: Key Takeaways about Plain Language from Healthwise

Healthwise is an organization that has embraced plain language. The organization devotes much of its time and energy to creating plain-language content—more perhaps than most organizations. Healthwise enacts several principles of ethical theory in its work, including respecting individuals' rights to make decisions about their own health care. Healthwise also maintains a feminist awareness of the differences in power between medical professionals and their patients—and uses plain-language information to address the disparity. Five lessons emerge from Healthwise's priorities and practices that could benefit other organizations:

1. **Support plain language from the top of the organization.** Plain language is an essential part of Healthwise's corporate mission. The CEO and other executives understand that they support the company's mission when they support plain language. Employees and customers alike know plain language is valuable when top executives support it. Healthwise executives publically support plain language through conference presentations, work with groups like the Center for Plain Language, and blog posts on the Healthwise website. They support plain language within the company by providing training, resources, and other support for plain-language work.

2. **Make plain language a part of the culture and the brand.** When plain language becomes part of the fabric of an organization, the entire organization will understand, appreciate, and support plain language. Everyone who works on content at Healthwise may suggest changes to make that content plain and clear for the target audience. Monthly norming meetings allow employees to share ideas and reflect on how to create effective plain-language content. Special events such as National Grammar Day and International Plain Language Day are opportunities to keep plain language in the organization's collective mind. Internal newsletters, shared plain-language resources, training sessions, and presentations at company meetings also help.

3. **Empower editors to enforce standards for plain language.** While Healthwise wants all content creators to use plain language, associate editors have job descriptions that require them to do so. Processes and policies both allow and require editors to edit content for plainness and clarity. Associate editors ensure that every piece of content the team writes is appropriately plain. Concern for plain language runs through the content-development process; it does not appear merely at the end of the process.

4. **Actively keep the audience in mind.** Employees at Healthwise understand that they are writing content for real people. Healthwise employees actively keep the audience in mind by using personas, by understanding the nature of complicated health and medical situations, by conducting usability testing, and by consulting frequently with experts who know the target audience well.

5. **Understand the ethical impacts of plain language.** At Healthwise, using plain language is one way of doing the right thing, doing right by those who read the company's content, and supporting the company's mission. Using plain language at Healthwise has as much to do with helping people as it does with attending to surface features like word counts, syllables per word, clear visual design, and sentence length.

Relevance of the BUROC Model at Healthwise

Healthwise recognizes that patients have rights to make decisions about the health and medical care they receive. Don Kemper founded the company on the premise that patients themselves are valuable resources in health care. Healthwise recognizes that many patients are unfamiliar with health-care bureaucracies and that medical jargon is often unfamiliar to patients. Decisions on health care, medical treatments, and lifestyle changes are critical, and Healthwise creates information to help patients make decisions with the best information available. The information Healthwise creates goes out to consumers through insurance companies and other large organizations as well as through internet resources and printed materials. Through its approach, Healthwise enacts what Obsorne calls the "ethics of simplicity" (2005, 51) by communicating health information in ways that are clear, simple, honest, and complete.

Healthwise executives understand the imbalance of power between medical professionals and patients. The company works to address and reduce this power differential in a manner consistent with feminist views of ethics (e.g., Walker 2001). Many Healthwise employees have current or previous experience as care providers, and they use that knowledge to shape the content they create. They help patients learn the language of professionals, and they bring the latest medical knowledge and research to bear on patients' concerns. In doing this, Healthwise provides a narrow ridge (Buber 1965) that helps patients and medical professionals hear and understand each other.

Healthwise uses a dialogic approach to content development through the three Ps approach, through writing with personas, and through the work of the UX team. Healthwise content cannot eliminate BUROC situations, but it helps patients make better decisions in response to them.

Questions and Exercises

1. In your organization, does support for plain language start at the top? If not, think about how you can influence the way top leaders view plain language. Identify the main concerns that your top leaders have about plain language, and create a plan of 300 to 500 words to address those concerns.
2. In your organization, does support for plain language come from many people or from only a few? Think about ways that you could promote plain language to others in your organization. Healthwise promotes plain language through internal newsletters, a plain-language glossary, training sessions, monthly meetings, and special events. What approaches might work for your organization? In a memo of 200 to 300 words, develop some ideas for promoting plain language in your organization.
3. How does your organization actively keep its audience in mind? Perhaps some aspects of the Healthwise approach might work for your organization.

On the other hand, perhaps some aspects of your organization's approach are especially effective already. First, identify the strategies your organization uses effectively and share them: blog about them, discuss them at a conference or an informal gathering, or share them in other ways. Second, identify some aspect you would like to improve, and then develop a plan of 300 to 500 words to improve it. Your plan should identify your improvement goal, the obstacles that currently impede your progress toward that goal, and the steps you will take to eliminate or work around the obstacles.

5

PROFILE–CIVIC DESIGN

Civic Design is not so much a company as an idea. Led by independent usability consultant Dana E. Chisnell, Civic Design works on projects at the intersection of information design, usability, and civic activity. The name Civic Design has appeared on projects since around 2011, but Chisnell has worked on civic design issues since 2001. Chisnell and other collaborators in Civic Design have completed several projects involving plain language in BUROC situations. In 2014, Chisnell and Whitney Quesenbery—also a usability expert and civic-design enthusiast with a strong interest in plain language—began to address similar projects through a new nonprofit organization, the Center for Civic Design (Dana Chisnell, pers. comm., November 19, 2013). In this chapter, I focus on how Civic Design produced a series of research-based instructional documents for county election officials: Field Guides to Ensuring Voter Intent.

This profile of Civic Design shows how a dedicated group of individuals worked without much funding to create an innovative set of plain-language documents that is free to county election officials across the US. Many people have goals to make important contributions to society, but they lack the resources. Civic Design shows how years of dedication to a cause and strategic relationship building can have a successful impact. This profile focuses on the highly bureaucratic and critically important process of voting and election management (which is common and yet unfamiliar to many, especially to new poll workers) that is oriented around citizens' rights to vote. It demonstrates how Civic Design supports ethical ideals of individual rights by working to ensure voters can cast ballots as they intend. It shows how Civic Design uses dialogue with county election officials to understand and address the challenges they face. It describes how Civic Design's content addresses BUROC situations, who creates plain-language content in Civic Design, how Civic Design creates its content, and how organizational culture and ethics affect Civic Design's work. The chapter concludes with lessons that plain-language professionals may take away from Civic Design.

Background of Civic Design

Chisnell started her career as a technical writer. After completing an internship with Burroughs, a maker of mainframe computers, and graduating from Michigan State University with an English degree, she worked at the Document Design Center at the American Institutes for Research in Washington, DC. Mentored by experts including Dr. Janice "Ginny" Redish, whose research appears in chapter 1, she learned principles of usability testing, plain language, and document design. Chisnell then spent more than a decade in several industry positions in technical writing. In 1999, she took a job with Tec-Ed, a US firm providing technical writing and usability-testing services. Chisnell worked in Tec-Ed's Silicon Valley office. Tec-Ed allowed her to shift her professional focus to usability. Since parting ways with Tec-Ed in late 2000, she has worked primarily in user research and interaction design (pers. comm., July 22, 2013). In recognition of her achievements in technical communication, her contributions to usability, and her public service, the Society for Technical Communication gave Chisnell its highest honor, the rank of Fellow.

Chisnell started Civic Design as a blog for all the writing and research she was doing about design and civic life and as a forum to announce activities such as speaking engagements and workshops. She volunteered her time for the vast majority of those civic-design activities and projects. In fact, many professionals have volunteered time on these projects and others like them. More recently, Chisnell obtained grant funding to support her work and her travel to promote the Field Guides at election meetings across the country.

Before starting the Civic Design blog, Chisnell had been involved in several other projects related to design and civic life. She participated in a project through the Usability Professionals Association (now the User Experience Professionals Association, or UXPA) called Usability in Civic Life. One major development of that project is the LEO Usability Testing Kit to help local election officials (or LEOs) to test ballots before voters use them in elections. Chisnell has also worked on a project with the American Institute for Graphic Arts (AIGA) called Design for Democracy, which views democratic processes through the lens of design. She also participated in development of the Anywhere Ballot, an accessible online tool that allows citizens to cast votes using any device with a standards-compliant browser. Because volunteers manage these projects without a lot of outside financial support, Chisnell said, many of the same people participate no matter what title appears on a project (pers. comm., July 22, 2013).

Chisnell has found strategic partners to support and promote Civic Design projects. One is Lawrence Norden, deputy director of the Democracy Program at the Brennan Center for Justice at the New York University School of Law. The Democracy Program focuses on voters' civil rights. Norden is an expert on voting rights, voting laws, and effective use of voting technology. Chisnell said Norden quickly understood the value of doing usability testing and the value of having plain language. Norden has helped arrange many usability-testing sessions on different ballots in several jurisdictions across the country; through his contacts, the Civic Design team has helped election officials in many states. Chisnell has also testified before the Presidential Commission on Election Administration, a non-partisan organization that works to identify best practices for elections and to improve the experience of voting for all citizens (pers. comm., July 22, 2013). Chisnell created a Kickstarter crowdfunding campaign to pay for designing and printing the first four volumes of the Field Guides. This campaign was so successful that it provided additional funding for research on additional Field Guides. Chisnell said the John D. and Catherine T. MacArthur Foundation supported research on volumes 5 through 8 of the Field Guides and funded her travel to promote the Field Guides at relevant conferences and events (pers. comm., July 26, 2013).

Motivation from Election Problems in 2000

Chisnell's interest in civic design issues developed after the US presidential election in 2000, in which Texas governor George W. Bush narrowly defeated Vice President Al Gore in the electoral college by winning the state of Florida's electoral votes. Chisnell said, "It was clear to me that the problems that went on in Florida were not technology problems, were not security problems, but were really usability and design problems" (pers. comm., July 22, 2013).

After months of recounts across Florida, votes cast in Palm Beach County swung the election in Bush's favor. Palm Beach County used the now infamous "butterfly ballot," in which the ballot holes corresponding to candidates in the presidential race appeared between two facing pages of candidate names. The confusing ballot "caused more than 2,000 Democratic voters to vote by mistake for Reform candidate Pat Buchanan, a number larger than George W. Bush's certified margin of victory in Florida" (Wand et al. 2001). In addition, statisticians Agresti and Presnell write, officials disqualified a relatively high number of ballots (4.2%) from the presidential election in Palm Beach County because of overvoting, or voting for more than one presidential candidate. Gore appeared on more than 15,000 of these ballots while Bush appeared on almost 4,000. "The evidence about the overvotes and the regression analyses of the Buchanan legal vote suggest that the butterfly ballot design is likely to have cost Gore a substantial number of votes in Palm Beach County. Given the small margin of victory statewide for Bush (537 votes), it is plausible that the ballot design cost Gore the election" (Agresti and Presnell 2002, 438).

Living in San Francisco during that time, Chisnell looked around for opportunities to learn how elections work. She found her way into a committee called the Ballot Simplification Committee for the city and county of San Francisco. California, like many other states, allows voters to decide the fate of many city, county, and state propositions. San Francisco County publishes a voter information pamphlet and distributes it to all registered voters before each election. These documents typically contain around 200 pages and provide a summary of each measure on parts of the ballot for the city and the county. The Ballot Simplification Committee has five communication experts who meet in public hearings to write these summaries in around 300 to 400 words. Chisnell said her experience working on the committee helped her see from the inside how the election division worked in her county. She found that the process for running elections is complicated and that county election officials face many tight deadlines that add to their challenges (pers. comm., July 22, 2013).

Civic Design Content and BUROC Situations

In a democratic society, voting for elected representatives is a critically important process. In the US, counties manage elections. Each county has its own bureaucracy to manage eligible citizens' access to voting. These organizations ensure that poll workers follow relevant processes appropriately; they ensure that citizens have information about elections and the contests within them. Although elections typically occur according to standard calendars, election workers must complete activities urgently on election days. Elections are often critically important: individuals elected to office can significantly affect public policy and legislation; referenda and propositions decided by the public can affect policies, budgets, schools, public facilities, and even relationships within communities. The right to vote is one of the great privileges of citizenship. The right to vote as you choose is fundamental to citizenship.

Chisnell noted that things have changed during the time that she and others have engaged in civic-design research. One such aspect is the nature of research on elections. When she first got involved, political scientists and social scientists were the primary researchers in that arena. Although these groups studied elections and found problems with the processes, they were not providing any solutions to those problems (pers. comm., July 22, 2013). Chisnell and her colleagues in usability and information design have applied their research and analysis to provide simple tools that local election officials can use. A second aspect is the type of work that county election officials must do. Chisnell said that county election officials are now responsible for managing information-technology systems more than ever before. Elections have shifted from paper punch-card ballots to optical scanning equipment and even electronic ballot systems. Yet these officials might not have backgrounds in either public administration or information technology (pers. comm., July 26, 2013). Because local election officials must not only oversee BUROC processes but also try to operate within them, they are likely to benefit from plain-language materials.

Personnel Who Create Plain-Language Content for Civic Design

With years of experience in technical writing and usability research, Chisnell did much of the writing for the Field Guides to Ensuring Voter Intent. That said, this project is another collaborative effort. One primary collaborator is usability expert Whitney Quesenbery, leader of Usability and Civic Life for the UXPA, a Fellow of the STC, and Chisnell's partner in the new Center for Civic Design. A second is designer Drew Davies, who leads AIGA's Design for Democracy initiative and has served as copresident of AIGA. Chisnell said that the overlapping projects of related organizations and the key activities within them—writing, usability testing, designing for pages and screens—share plain language as the common thread: "All of that is plain language. All of it includes plain language. Every single bit of it, whether we call it that specifically or not" (pers. comm., July 22, 2013).

In addition to writing, the Civic Design team needed to apply many other skills to create and publish the Field Guides. Each volume of the Field Guides lists members of the team who helped create it. Other team members include advisors, a strategist, a videographer, a public-relations specialist, and an illustrator.

Practices and Processes for Creating Plain-Language Content

Through Civic Design, Chisnell and her collaborators have used several strategies for creating effective plain-language content. These include participating actively in the audience's domain to understand the audience's needs, using existing knowledge to develop content, collaborating with a wide variety of like-minded people, letting design decisions influence how they write, and challenging conventional wisdom.

Participate Actively in the Audience's Domain to Understand the Audience's Needs

Chisnell and her collaborators in Civic Design learned to understand civic-design problems before trying to solve them. Civic Design team members have learned much about the audience for the Field Guides through years of participating in civic processes and conducting workshops and conference sessions for local election officials.

Learning before trying to lead is critical. Chisnell said, "We have learned through the years that we can't just show up and tell them [election officials] that we know how to do their jobs, because we don't" (pers. comm., July 22, 2013). Chisnell's experience on the Ballot Simplification Committee helped her understand how time complicates and constrains the election process. She learned how officials create important documents such as voter-education materials and poll-worker training materials. She learned that in her county, the person in charge of creating paper ballots did not personally design the ballot; instead, this person gave information to the voting-system vendor, who then laid out the ballot and sent it directly to the printer without getting feedback. Describing some of the lessons she had learned, Chisnell said,

> These are things I didn't have insight to before. Learning them and having a better understanding of what the process was gave me some knowledge about where the pain points might be, and where things like the butterfly ballot might happen. Now, it turns out that the butterfly ballot is probably the most famous but not necessarily the most egregious thing that has happened to a ballot. And, Florida is not alone with its style and design problems; it happens all over the place. (pers. comm., July 22, 2013)

Over several years, Civic Design has helped local jurisdictions recast instructions on forms related to voting; it has gathered volunteers to run "flash" usability tests (informal tests with nonrandom convenience samples of willing participants) on forms or ballot designs over the course of a Saturday, and it has conducted workshops on usability and clear communication. Chisnell said the team gradually started to promote the approach described in the Field Guides through presentations and workshops (pers. comm., July 26, 2013).

Another aspect of Civic Design's active participation in civic processes is the team's desire to improve the experience of voting and ensure that voters can vote as they intend. Chisnell said, with a chuckle, that a central advantage that her team offers is that they are not university-based academics. She said that political scientists, social scientists, and even computer scientists study elections frequently but that local election officials get little practical value out of the research findings (pers. comm., July 26, 2013). By writing and producing the Field Guides, Civic Design demonstrated its desire for elections to go well for election officials and voters.

Over the years, Chisnell has learned much about local election administrators and the challenges they face. Chisnell offered this composite description of a typical local election administrator:

> Depending on where you live, that person might be the county clerk or the county registrar. He or she—probably a she who has been in her job for about 20 years or so—might be elected, might be appointed, and has another job besides just running elections. Often they are the people in the county office who manage vital records: birth certificates, death certificates. They might also handle deeds and issue licenses; basically, they're recorders. In larger counties, the person who is the election official is not necessarily the clerk. But that official probably was hired or appointed by the clerk or the registrar. And over the last ten years, running elections has become more and more of an IT job than practically anything else. So, these people are public administrators but often don't have a degree in public administration. They come from a whole bunch of different

backgrounds. They might even have done something else completely different in one career, retired from it already, and moved to public service in their county. They are awesome public administrators, but they are not designers. (pers. comm., July 26, 2013)

It is not surprising that election officials, similar to this one Chisnell described, would keep ballot designs, voter instructions, and other documents the same from year to year. These election officials know how to do their jobs well. It is likely, however, that without the help of specialists like those in Civic Design, they would not get insights into ballot design or how to ensure voter intent. Pressed for time and lacking resources, election officials will likely make only a few changes to their ballots and documents, if any. Because Civic Design has built relationships with election officials over years of participating in election administration and sharing knowledge in presentations and training sessions, the group deeply understands the design possibilities available for local election administrators.

Use Existing Knowledge to Develop Content

The practice of using existing knowledge to develop content is not new. Plain-language professionals commonly conduct research to find the information that their audiences need. That said, the Field Guides appear to be the first to take existing material from extensive, thorough research reports and present it in ways that are easy for local election officials to use.

Chisnell said, "There were these big juicy reports sitting out there—two- and three-hundred page reports—that had beautiful specifications and guidelines in them, but they were totally inaccessible to people who are really busy doing these other things.... They were very much about where the agency was that was sponsoring them, the research, and not very much really about who election officials were" (pers. comm., July 26, 2013). These documents included a collection of best practices in design produced by the US Election Assistance Commission and two items from the National Institute of Standards and Technology: a style guide for voting system documentation and a report on effective ballot language.

Discouraged that "there were these rich resources gathering bit-dust on them out on the Web," Chisnell and Civic Design dived into the reports and extracted information that they thought would be most useful (pers. comm., July 26, 2013). The first four volumes of the Field Guides include information from this extensive body of research.

Collaborate with Like-Minded People—No Matter Who or Where They Are

Chisnell started Civic Design as an idea as much as an identity; the goal was to accomplish the work of civic designing, not to promote herself or her own business as an independent contractor. Civic Design team members have a variety of educational and work experiences, and they live across the US. Over the years, Chisnell has contributed to projects with Usability in Civic Life and Design for Democracy, and members of those projects have participated in Civic Design projects such as the Field Guides. Connecting with the Brennan Center for Justice has allowed Civic Design to meet like-minded researchers and advocates while also helping more officials who manage elections.

In several instances, Chisnell has used networks within professional organizations to enlist volunteers. For example, volunteers helped improve voting practices in Minnesota. In 2008, Minnesota had a contentious election for a US Senate seat. Al Franken, a Democrat, and Dean Barkley, an independent, challenged Republican incumbent Norm Coleman. After the first ballot count, Coleman led Franken by several hundred votes. As local election officials rechecked their figures, Coleman's lead dwindled to 215. The narrow margin triggered a mandatory recount of ballots by hand. After many legal challenges and court reviews, Franken eventually won by 312 votes (Foley 2011). Many ballots that the two candidates' campaigns challenged for validity in the recount were absentee ballots submitted by mail. The Minnesota secretary of state's office reached out to Chisnell for help with usability testing of the absentee ballot. Chisnell was not able to attend herself, but she and Quesenbery called upon technical writers, usability specialists, and technical illustrators in the Minneapolis–St. Paul area. These professionals volunteered to work with the secretary of state's staff to conduct usability tests in a branch of the Saint Paul Public Library. This successful round of testing led to a second set of usability tests. Chisnell said Secretary of State Mark Ritchie so appreciated the improvements gained through usability testing that he became an advocate for plain language (pers. comm., July 22, 2013). In 2010, the Center for Plain Language recognized Minnesota for its effective plain language, naming Minnesota's instructions for voting with an absentee ballot as a finalist for its ClearMark award for best revised document in the public sector.

Let Design Decisions Shape Content

Through their design, the form and function of the Field Guides complement each other effectively. Each Field Guide is a small booklet that, when closed, is about the size of an index card. The design is at once utilitarian and stylish. The covers are made of durable brown paper stock that resembles a sack for groceries, and the contents are in black ink on white paper. At the same time, the corners are round to add a bit of flair. Chisnell laughed when she called the Field Guides "adorable because they're small." Civic Design selected Scout Books in Portland, Oregon, to publish the Field Guides.

As Civic Design developed its vision for the Field Guides, Chisnell and her collaborators asked what advice they would give to election officials if they could

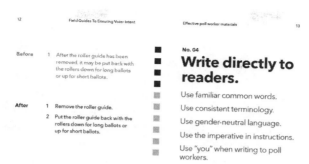

FIGURE 5.1 Two-page spread from "Field Guide, Volume 4" on effective poll-worker materials. Reprinted with permission from Dana E. Chisnell.

tell them only a few things. Chisnell knew that making even a few small changes could not only help ensure voter intent but also help improve the efficiency of election officials' processes. She added that the compact design for the Field Guides was "a good constraint for us because we didn't get carried away writing long explanations of these things that would, you know, bog people down" (pers. comm., July 26, 2013). Figure 5.1 shows a succinct two-page spread from volume 4 about writing to poll workers.

Each Field Guide provides 10 guidelines for practice in an aspect of conducting elections. These suggestions are direct and prescriptive, with examples and explanations that support best practices. Figure 5.1 shows a suggestion, "Write directly to readers," followed by five specific examples of how to do that: use familiar common words, use consistent terminology, use gender-neutral language, use the imperative in instructions, and use "you" when writing to poll workers. On the left page, readers see examples of how to apply the suggestion in black ink, along with negative examples of what not to do in gray ink. In the center, readers see a column of 10 squares. The number of bold black squares on each spread identifies the number of the suggestion. Chisnell added that each guideline takes into account the realities facing election officials: time constraints, budget constraints, laws and regulations, and technological constraints.

Each Field Guide ends with a checklist to help election officials know when they have understood the guidelines appropriately. Said Chisnell, "One of the problems often with guidelines is that you can talk about them all you want, but how do you know if you're actually following them? So we put a little quiz, a little test, at the end of each of the Field Guides to say, 'If you're doing these things, you're getting it right'" (pers. comm., July 26, 2013). A straightforward quiz is a learning technique that students encounter frequently. Promoters of health literacy often encourage care providers to have patients "teach back" what they have learned so that providers can gauge patients' understanding.

Challenge Conventional Wisdom

Challenging conventional wisdom relates to the practice of letting design decisions shape the content. Figure 5.1 shows a two-page spread from a Field Guide. The guideline is on the right page while the example showing how to apply the guideline is on the left. Because Western readers read from left to right and top to bottom, many communicators would put the guideline on the left and the examples on the right. Chisnell said that the Civic Design team decided to challenge that practice, and the results from readers have been encouraging:

> We know that people look at the right-hand page first. They always do and they've been trained to do that in a variety of ways. But part of it is just the ergonomics of using a book. You open the cover and the first page you see is the right-hand page. So we put the guideline on the right and the examples on the left. And that seems to work really, really well. (pers. comm., July 26, 2013)

Figure 5.1 gives an example of this design. While a longer book with many more pages might be difficult to read by starting on the right side, the arrangement works well enough for readers of the Field Guides and contributes to their stylish, even quirky charm.

This is not Chisnell's first time to challenge conventional wisdom when working on a civic-design issue. She gave another example from her work on the Anywhere Ballot, the flexible front-end tool available for accessible online voting on a variety of electronic platforms. Including an illustration in the instructions for marking a ballot in an election is common. But Chisnell and colleagues found that illustrations confused some voters with low literacy and cognitive problems: "People expected to be able to interact with the illustration because of [the technology] they were using it on, so we ultimately took the illustration away. It wasn't really needed" (pers. comm., July 26, 2013). Users of browsers often click on images to expand them and to zoom in on small details. The designers of the Anywhere Ballot expected users to treat pictures as mere illustrations, but the actions of the test subjects showed that they had different expectations. In many cases, the conventional wisdom turns out to be less valuable than people expect.

Another earlier example is the LEO Usability Testing Kit developed for local election officials. As Chisnell described it, "This was heretical at the time, because we were simplifying usability testing to the extent that somebody who had no training could go observe voters interacting with the ballot, see where there were issues, and make some new design decisions out of that" (pers. comm., July 22, 2013). Some in the usability community opposed efforts to develop such a kit because it would take away opportunities for consultants to obtain paid work. Others said that local election officials would not use the kit if they had it.

Chisnell acknowledged that local election officials do not have much time to do usability testing, but she still believes that the LEO Usability Testing Kit has been a valuable tool:

> It's not being done much because there are so many deadline pressures. Just fitting in the time to have a day to do it when what you're focused on is getting the ballot to the press or getting the [ballot] programming done, because you only have five days to do it, is tough. But we've demonstrated usability testing at a bunch of conferences where there are election officials and political scientists and academicians, so I believe that more people will be doing it themselves over time. It's sort of an evolutionary thing: "Show us the difference that good design can make, then show us the technique you used to get that, and then show us how to do that technique." And we're kind of between the second and third steps right now, I think. Which I'm delighted about, because there aren't enough of us [laughing]. (pers. comm., July 22, 2013)

As more election officials learn about usability testing and document design, they will be able to share and develop expertise among themselves. As usability and design knowledge start to reside within the community of election officials, perhaps that knowledge will spread more rapidly than it does when shared by groups like Civic Design.

Organizational Culture and Plain Language in Civic Design

The tag line on the Civic Design home page summarizes the group's point of view: "Democracy is a design problem" (Chisnell 2013). Team members analyze election issues as designers—of documents, of interfaces, of user experiences. To echo a previously noted comment from Chisnell, all of that includes plain language, whether labeled "plain language" or not.

Plain language is a means of addressing design issues in civic activities, but Civic Design does not promote plain language as an end in itself. Instead, plain language is a means to better inform voters and to help them vote as they intend. Civic Design team members uphold ideals from civic society and from the realm of design practice. In a representative democracy, ideals include individuals' right to vote—regardless of race, gender, ethnicity, socioeconomic status, or any other characteristic used as a reason to discriminate—and their right to vote as they intend. Two design ideals are ease of use and accessibility to broad audiences. Thus, Civic Design team members uphold ideals from civic society and from the realm of design practice.

Projects such as the AIGA's Design for Democracy and the UXPA's Usability in Civic Life persist primarily through the work of volunteers. As mentioned earlier, Chisnell started the Civic Design website as an outgrowth of her work as a volunteer. These volunteers participate actively in civic life by working to help other citizens—regardless of political affiliations or ideologies—exercise their rights as voters. The Field Guides support the ideal that every voter's vote counts, literally and figuratively.

Chisnell said that ethics influences both why and how she completes projects through Civic Design. In explaining why she works on Civic Design projects, she mentioned the example that her parents set. They were, she said, always active in the community, and they regularly exercised their right to vote. Chisnell's mother was active in millage campaigns. States such as Michigan, where Chisnell grew up, fund education through millage. (A mill represents one dollar per every thousand dollars of a home's value, and states levy school taxes at a certain number of mills.) Moving into technical writing and, later, into user research and usability testing, Chisnell said she had a sense of social responsibility and of trying to make the world a better place through her work. Creating the Field Guides to Ensuring Voter Intent has been part of her desire to do ethical work (pers. comm., July 26, 2013).

In addition to doing work that benefits society at large, Chisnell strives to conduct her work and respect the human test subjects with whom she works in ethical ways. Chisnell has submitted several research protocols to institutional review boards over the years, so she knows principles for treating human research subjects ethically. (Institutional review boards in universities, hospitals, and other organizations that do research inspect research protocols to ensure that researchers treat all of their test subjects humanely and with respect. They can disallow any research proposals that do not meet ethical standards.) She has completed online research ethics training from the US Department of Health and Human Services, and she frequently recommends it to other colleagues in user research. Chisnell notes that while researchers connected to US colleges and universities are required to complete ethics training, those who conduct user research in the private sector typically are not (pers. comm., July 26, 2013). Although some might believe that the ultimate ends of a certain type of testing might justify the means, Chisnell keeps concerns about her human test subjects paramount. Chisnell demonstrates sincere concern for her test subjects by completing training in research ethics even when she has no professional requirement to do so.

For example, Chisnell notes that in online forums for usability researchers, she has seen questions about whether it is acceptable to give test subjects a task that is impossible to complete successfully. She notes that often the desire to create such a test comes in response to an issue involving politics or dissenting views between members of a team. Fortunately, Chisnell said, "the community standard seems to have remained constant. That is, don't do that except under extremely specific circumstances where there's a lot of support for the user" (pers. comm., July 26, 2013). Appropriate support for test subjects includes a thorough description of the test procedures, a test administrator who monitors participants closely, and an opportunity for test participants to quit if they do not like how the test is going. Chisnell emphasized that it is important to recognize and reinforce the rights of participants in usability testing: "If you see study participants as data points, it's easy to make that abstraction and do the equivalent of injecting them with bad things or making them do things in sessions that they wouldn't normally do" (pers. comm., July 26, 2013). A key to ethical research with human subjects is to provide them enough information so that they may give informed consent before participating. It is unethical to mislead test participants or to assume without having asked them that they want to participate in the research.

In considering Buber's I–It mode of communication (speaking at someone) and the I–You mode (in which interlocutors demonstrate mutual respect), Chisnell said that the Field Guides "absolutely" fit the I–You model: "We often do exercises with election officials about how they perceive instructions that they get under the I–It model versus the I–You model. And one of the things that inevitably becomes apparent is—like passive voice problems—they don't know who the *It* is. Who? How? Who does that actually affect? Is this me, really?" (pers. comm., July 26, 2013). Election officials are experts, and it can be difficult for them to understand elections from a nonexpert's perspective. By helping election officials see the voting experience from their constituents' perspectives, Civic Design team members help election officials to understand their audiences and to strive for dialogic, respectful communication instead of monologic and coercive communication.

Chisnell said that local election officials deal with dense, turgid regulations all the time; their experiences sometimes help them understand how their constituents feel about reading convoluted documents. Chisnell added that some officials have seen good results from having conversations with their constituents rather than simply talking at them. These officials understand "that I–You is going to be more successful for them" because they have already seen the benefits firsthand (pers. comm., July 26, 2013).

Conclusion: Key Takeaways about Plain Language from Civic Design

Civic Design is at once the vision of one person, Dana Chisnell, and the fruit of the efforts of several people. The ethical principles reflected in Civic Design's work include strong respect for citizens' rights, the obligations of election officials to help citizens exercise their rights, and support of ethical values such as justice and fairness. The efforts of Civic Design appear in usable ballots and improved election processes in counties across the US. These lessons from Civic Design might benefit other organizations as well:

1.   **Use plain language as a means, not the end**. Civic Design addresses democracy as a design challenge. Ballots, forms, voter information pamphlets, websites for county election offices, poll-worker training materials—all

these and more provide practical opportunities to communicate effectively with large, diverse audiences. Civic Design team members approach election processes holistically; they do not focus on only one area. Whether writing content for Field Guides, summarizing results of a usability test, or helping develop an online voting tool, Chisnell and her collaborators apply the principles of plain language to do their jobs effectively.

2. **Invest the time needed to understand practices thoroughly.** Chisnell started learning about election processes a decade before writing the first volumes of the Field Guides to Ensuring Voter Intent. She realized that if the problems around elections were simple and easily addressed, someone would have solved them already. Having spent so much time learning about elections as volunteers, Chisnell and her collaborators now speak as experts. This expert status allows them to influence more election officials and to improve more documents and systems that voters use.

3. **Use existing research to improve practice.** Extensive reports on election practices and ballot design have been available for several years, but county election officials lack the time and the experience to extract useful information from them effectively. Civic Design used these resources to give valuable and effective advice to county election officials—in plain language.

4. **Be willing to challenge conventional wisdom.** Civic Design has taken some unconventional approaches to developing successful projects. Chisnell took a somewhat heretical stance when helping to create the LEO Usability Testing Kit for election officials who lack training and experience in user research. When designing the Field Guides in two-page spreads, Civic Design violated left-to-right reading patterns by placing guidelines on the right-hand page and supporting text on the left. Instead of waiting for clients to fund their work, Civic Design members invested their time without pay; crowdsourced funding and a foundation grant later brought Civic Design a new level of exposure and influence.

5. **Collaborate with like-minded people.** Partners in Civic Design do not worry much about what organization's name will appear on a project. Instead, they work on projects in which they believe. They work with people whose experiences will complement their own. In some cases, such as usability-testing ballots across the country, they have empowered knowledgeable volunteers to participate in their work. A shared respect for the principles of plain language and a focus on the rights of every voter have allowed Civic Design to influence election practices in the US for the better.

The BUROC model identifies challenges that face election officials who oversee voting and citizens who want to cast their votes. Voting is a bureaucratic activity, both for citizens and for election officials. Election officials must understand policies and procedures for registering and for casting votes, and they must convey them to their constituents; citizens must understand those policies and follow them. Elections occur sporadically, some citizens might not vote regularly, and other citizens might move to new counties or states with unfamiliar procedures to follow. Election officials must hire poll workers, many of whom will not be familiar with voting processes and regulations. The right to vote is one that many citizens cherish. Processes that help voters correctly express their intent in the voting booth enact the ethical principle of utility or usefulness while they also fulfill the Kantian obligation to treat people as ends and not merely means (Kant [1785] 1969). Voting to elect officials is a critical process with tight and strict deadlines, and individual votes are truly important—as the 2000 US presidential election and other close elections have shown. Elections require the involvement of many people, and yet election officials often lack the time and resources to train election workers extensively. As Civic Design has shown, plain language is an effective tool to use to address BUROC situations.

Questions and Exercises

1. Dana Chisnell uses many personal and professional networks to complete Civic Design projects. How well is your organization connected to other similar organizations? Does it have a good network of organizations within the same industry or perhaps the same region of the country? In 150 words or so, write down or map out the primary connections your organization has to others. Evaluate these connections: Is this list extensive? Does the list show some diversity? Are some connections old, some new? Are some connections larger, some smaller? If you are not satisfied with the state of the list, identify ways to add more connections to your organization's network.

2. Identify the documents or genres that your organization uses to communicate with its constituents. Get creative; think about how you could challenge the conventional wisdom around one of these documents to better reach your audience. What if you radically redesigned it to fit on a business card or a bumper sticker? What if you turned it into a checklist? Think about what your organization really wants the audience to do with this document, and consider whether a change in its form might make it more usable. Write up the ideas you generate in 200 words or more.

6

PROFILE–RESTYLING THE FEDERAL RULES OF EVIDENCE

The Federal Rules of Evidence govern the ways that litigants may enter evidence into proceedings of a US federal court. Because most states model their evidence rules on the Federal Rules of Evidence (or the Evidence Rules), these federal rules affect judges, attorneys, plaintiffs, and defendants across the US.

A group called the Standing Committee on Rules of Practice and Procedure (or the Standing Committee) maintains and reviews the rules governing US courts. These rules affect five areas of legal procedures: civil courts, criminal courts, appellate courts, bankruptcy courts, and evidence. The Advisory Committee for each of the five areas reports to the Standing Committee. From 2007 to 2011, a group of judges, law professors, and other legal experts on the Advisory Committee for Evidence Rules updated the Evidence Rules by restyling them in plain language whenever possible. The Standing Committee calls this process "restyling" because the goal is to change the style of the documents without changing the substance. Many of the people who participated in restyling the Evidence Rules gathered for a symposium in October 2011. The published record of that symposium (Douglas et al. 2012) provides a wealth of insights into the processes used in restyling and the challenges of making stylistic changes to a document that thousands of legal professionals use.

I profile the restyling of the Evidence Rules because the process shows how people worked together to create a successful plain-language document in an environment—the legal profession—in which people are suspicious of or even outwardly hostile toward plain language. Kimble (2006, 2012) writes that some legal professionals promote legalese over plain language because, among other reasons, they think that plain language is imprecise, that it lowers professional standards, or that it will reduce the amount of billable work available. Suffice it to say that Kimble rebuts all of those arguments and more. The awards the restyling project received also show that it is noteworthy. The restyled Evidence Rules won the 2011 Burton Award for Reform in Law from the Burton Foundation, which honors outstanding legal writing, and a 2011 ClearMark Award from the Center for Plain Language (Douglas et al. 2012, 1438).

This chapter describes how participants in the restyling project updated the Evidence Rules. It also describes how the Evidence Rules address BUROC situations, how participants completed the restyling project, and how organizational culture and ethics affected the Evidence Rules project. It closes with lessons that plain-language professionals may take away from the effort to restyle the Evidence Rules.

Background of the Project to Restyle the Federal Rules of Evidence

Judge Robert Keeton, who led the Standing Committee in the early 1990s, led early efforts to make the rules for federal courts easier to read, understand, and follow. Figure 6.1 shows the hierarchy of the Standing Committee and the subcommittees that oversee federal rules of judicial practice.

According to Judge Sidney A. Fitzwater, it was Keeton and Professor Charles Alan Wright who led an effort to adopt clear and consistent style conventions for all the national rules of procedure:

> The rules had been enacted without consistent style conventions, so there were differences from one set of rules to another, and even from one rule to another within the same set. Different rules expressed the same thought in different ways leading to a risk that they would be interpreted differently. Different rules sometimes used the same word or phrase to mean different things, again leading to a risk of misinterpretation. And drafters made no effort to write the rules in plain English. (Douglas et al. 2012, 1438)

If it is possible for one court to interpret a rule differently than another court does, then it is difficult to ensure that courts will be fair to all participants. As Fitzwater describes them, rules in plain language are easier for all courts to understand, interpret, and apply consistently.

Judge Keeton created a Style Subcommittee of the Standing Committee and enlisted legal writing expert Bryan A. Garner as the first style consultant to the subcommittee. Garner led restyling work on the Federal Rules of Appellate Procedure, which the Supreme Court approved in 1998, as well as on the Criminal Procedure Rules, approved in 2002. Garner also created a document to guide restyling efforts, "Guidelines for Drafting and Editing Court Rules" (Garner 2007). Professor Joseph Kimble took Garner's place in 1999, after the work on Criminal Procedure had finished (Kimble, pers. comm., October 23, 2013). Kimble led work on the Civil Procedure Rules, approved in 2007, before starting on the Evidence Rules in late 2007 (Kimble 2009, 46). By creating and supporting the Style Subcommittee, Judge Keeton prepared the way for the restyling of the federal rules.

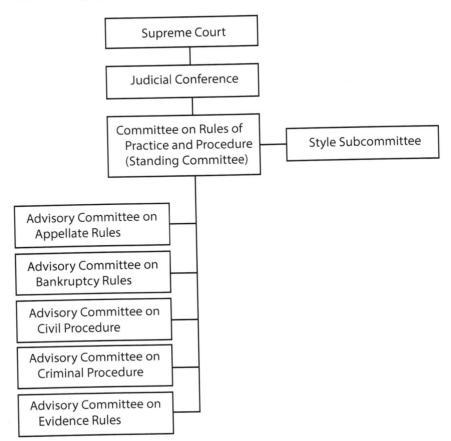

FIGURE 6.1 The Standing Committee and its associated committees. Adapted from Kimble (2013).

According to Kimble, the old Evidence Rules were "riddled with inconsistencies, ambiguities, disorganization, poor formatting, clumps of unbroken text, uninformative headings, unwieldy sentences, verbosity, repetition, abstractitis, unnecessary cross references, multiple negatives, inflated diction, and legalese" (Kimble 2009, 46). Because of their many shortcomings, Kimble called the Evidence Rules "a professional embarrassment" (46). Kimble's critiques show that the old Evidence Rules were not only hard to read but also hard to use. A difficult, unclear document does not reflect well on the organization that produced it.

The Evidence Rules are truly important, as are all of the court rules of practice. According to Judge Geraldine Soat Brown, more than 1,000 federal judges use the Evidence Rules (Douglas et al. 2012, 1475). Professor Paula Hannaford-Agor of the National Center for State Courts pointed out that 41 states and Puerto Rico have evidentiary rules that substantially overlap with the federal Evidence Rules

(Douglas et al. 2012, 1514). The Evidence Rules help the courts do their work. According to Judge Marilyn L. Huff, "the restyling project benefits the bench, the bar, and the public by improving clarity and consistency. When the Rules are clear, litigation over ambiguous meaning decreases and compliance with the Rules increases. The net result is an excellent example of the rules enabling process at work to improve the administration of justice" (Douglas et al. 2012, 1464). Thus, the restyling of the Evidence Rules (and all the other rules overseen by the Standing Committee) is not merely an exercise in substituting clear language for unclear language. The restyling efforts help all people working in the courts to understand—and thus, to meet—their responsibilities and opportunities more effectively.

Federal Rules of Evidence and BUROC Situations

Court proceedings at the federal and state levels certainly qualify as BUROC situations. The Evidence Rules affect what attorneys can do within the legal bureaucracy to protect their clients' interests. Kimble described how court trials affect individuals and society as a whole:

> Your freedom can depend on it. Your life can be affected in almost every possible way. The public good can be affected by lawsuits, often dramatically. (Think of decisions by the US Supreme Court—in fact, by any appellate court and even by trial courts.) People usually go to court only as a last resort, and on important matters—important personally or publicly or both. (pers. comm., October 23, 2013)

As the courts make important decisions on guilt or liability, or the absence thereof, the results can be urgent and critical, and they may uphold or suspend individuals' rights as the law determines.

In the symposium, Judge Robert A. Hinkle used a football analogy to describe how a trial is a critical event in which lawyers and judges must react urgently. A fan of Florida State, his alma mater, Hinkle recalled listening as a student to a radio call-in show featuring the team's coach, Bobby Bowden:

> Invariably someone would call in and point out some play Coach Bowden should have run in the prior Saturday's game that would have been better. Coach Bowden would say something like, "Yeah, buddy, you're right, but dadgummit, I only had twenty-five seconds. You had three days to come up with that."
>
> If you are the lawyer trying a case, you don't have twenty-five seconds to decide whether to object. You don't even have five seconds. If you wait five seconds after your opponent asks a question, the witness is going to blurt out the answer, and the judge—if the judge is a little too

clever—may say something like, "If you chance to hear the answer, you waive the objection."

And so you don't have five seconds. As a judge, if I pause to think for five seconds, the witness will blurt out the answer, or a lawyer will start talking—making a speaking objection or speaking response, even though I've told the lawyers not to do that. So a judge really has to be on top of this. (Douglas et al. 2012, 1444–45)

Here, Hinkle provides the trial judge's perspective on critical events in a trial. These events are critical for plaintiffs, defendants, and their attorneys as well. Nor are these critical, urgent events isolated. According to Judge Joan N. Ericksen, a trial judge might have to make 30 to 40 rulings on evidence in a single day of trial (Douglas et al. 2012, 1460). Ericksen further underscored all that is at stake in a court trial:

Trials are expensive. They are high-drama, high-stakes events. The clients are terrified. Lawyers lose sleep. They forget to make their mortgage payments. They ignore their families. It takes years to develop the great courtroom skills of organization and persuasion and facility with the Rules of Evidence, which, after all, are the rules that control it all. (Douglas et al. 2012, 1461)

As these descriptions from Judge Ericksen, Judge Hinkle, and Kimble show, court proceedings are BUROC situations in which the participants affected may benefit from plain-language documents that they can understand and apply without arduous effort.

Although attorneys and judges are familiar with court rules and policies, plaintiffs and defendants who choose to represent themselves pro se in court are often unfamiliar with them. According to Professor Hannaford-Agor, state courts conduct many more trials than do federal courts, and state courts see more pro se litigants. Professor Hannaford-Agor said that because the federal Evidence Rules influence so many states' rules, the restyled Evidence Rules have potential to benefit state courts substantially along with the federal courts (Douglas et al. 2012, 1513).

Personnel Who Create Plain-Language Content

The project to restyle the Evidence Rules involved many people on the Standing Committee, the Style Subcommittee, and the Advisory Committee on Evidence Rules. Individuals on these committees serve unpaid appointments from the Chief Justice of the Supreme Court. Three who were most actively involved include Judge Hinkle, chair of the Advisory Committee on Evidence Rules during the restyling project; Kimble, the style consultant on the Style Subcommittee; and

Professor Daniel J. Capra of Fordham Law School, the reporter for the Advisory Committee on Evidence Rules. Judge Hinkle served as a trial lawyer for 20 years before becoming a federal district judge (pers. comm., November 28, 2013). Kimble taught for decades at Western Michigan University Thomas Cooley Law School in Lansing, Michigan; he has also served as editor of *Scribes Journal of Legal Writing* and the "Plain Language" column of the *Michigan Bar Journal* (Douglas et al. 2012, 1439–40), in addition to writing books on plain language. Fitzwater, who succeeded Hinkle as chair of the Evidence Rules committee, called Capra's knowledge of the Evidence Rules "encyclopedic," and he called Kimble and Capra "the sine qua non of the restyled Evidence Rules" (Douglas et al. 2012, 1439).

Many members of the Standing Committee, the Style Subcommittee, and the Advisory Committee on Evidence Rules spoke at the symposium on the restyled evidence rules or submitted remarks for the record (Douglas et al. 2012). These individuals included judges, attorneys, and law professors.

Practices and Processes for Creating Plain-Language Content

The project to restyle the Evidence Rules used several practices that led to a successful conclusion. These include establishing a clear restyling process, clearly defining the scope of changes in style, leaving certain textual problems unchanged to protect usability and audience expectations, seeking feedback and considering it carefully, and using style guides for reference and support.

Establishing a Process with Clear Lines of Authority

Kimble described the restyling process this way: As the plain-language consultant, he produced the first draft of each revised rule. He sent each draft to Capra, the reporter for the Advisory Committee. When Capra sent back comments, Kimble revised in light of them. Judge Hinkle added that sometimes a draft cycled between Kimble and Capra more than once (pers. comm., November 28, 2013). Unresolved issues went to the Style Subcommittee. Kimble revised again in light of the Style Subcommittee's decisions and suggestions, and then the draft went to the Advisory Committee for a full review (Kimble, pers. comm., December 1, 2013).

Members of the restyling project did their work via email, telephone, and occasional face-to-face meetings. Hinkle, chair of the Advisory Committee during restyling, said that the process was somewhat flexible and not rigid in every instance. Yet the process of separating style concerns from substance concerns remained consistent: "The final decision was the Style Subcommittee's on matters of style, and the Advisory Committee's on matters of substance. The Advisory Committee had the final say on whether a change dealt with substance or only style" (pers. comm., November 28, 2013). At the symposium, Capra called the work of the Style Subcommittee "quite remarkable" (Douglas et al. 2012, 1461). The group's work was remarkable because project members completed a large amount of work over the course of three to four years and because the project was particularly challenging and labor intensive. Another remarkable aspect of the project is that group members had many discussions and often disagreements, but their professional relationships stayed intact. Kimble (2013) said that the experience was challenging and painstaking yet always collegial and focused on the project's benefits to the profession.

Defining the Scope of the Style Changes

The practice of defining the scope of style changes relates closely to the practice of establishing clear lines of authority. The Standing Committee called the process "restyling" to emphasize that participants were not changing the substance of the rules. Kimble pointed out that a separate process exists for changing the substance or content of rules and procedures (pers. comm., October 23, 2013). Citing Hinkle, his predecessor as the chair of the Advisory Committee, Fitzwater said that both the old rules and the restyled rules should lead a lawyer or judge to the same result, but the restyled rules should reduce the possibility of misunderstandings between participants in a trial (Douglas et al. 2012, 1440). Participants in the project developed principles for determining which changes to allow and which to prevent. Fitzwater said project guidelines prohibited substantive changes and changes to especially familiar phrases:

> A proposed change was substantive under any of these circumstances: under the existing practice in any circuit, it could lead to a different result on a question of admissibility; under the existing practice in any circuit, it could lead to a change in the procedure by which an admissibility decision is made; it changes the structure of a rule or method of analysis in a manner that would fundamentally alter how courts and litigants have thought about or argued about the rule; or it changes what has been referred to [by Judge Hinkle] as a "sacred phrase," a phrase that has become "so familiar as to be fixed in cement." (Douglas et al. 2012, 1440)

Kimble said that on some occasions, the committees did have to vote on whether a particular change was substantive or purely stylistic. He also pointed out that an archive with several hundred documents records discussions and comments made throughout the project (pers. comm., October 23, 2013). While it can be difficult to distinguish between a text's style and its meaning, the definitions established by project participants show how they focused the scope of their work.

Professor Katharine Traylor Schaffzin of the University of Memphis added that the sacred phrases are those so familiar in practice that to alter them would create undue disruption (Douglas et al. 2012, 1491–92). Schaffzin compared the old Evidence Rules with the restyled rules, and she judged that the process of determining sacred from not must have been arbitrary. Andrew D. Hurwitz, then vice chief justice of the Arizona Supreme Court, explained with a bit of hyperbole and humor that the process was not arbitrary:

> The way it worked was that virtually every member of the Committee, largely led by Judge Ericksen, believed *every* phrase in the old Rules was sacred. Professor Kimble believed that *no* phrase in the old Rules was sacred. And therefore on each issue, as we normally did on technical evidentiary matters, we ended up relying on Professor Capra to tell us which ones were sacred and which ones were not, and that worked out pretty well, I think. So whether or not the system was perfect, it wasn't arbitrary at all. It was quite predictable. (Douglas et al. 2012, 1502; emphasis in the original)

All joking aside, Hurwitz's comment shows how committee members worked systematically and provided checks and balances on each other's understandings of the style of the Evidence Rules.

Protecting Usability and Audience Expectations

Members of the restyling project were aware that changes to the Evidence Rules could disrupt the work of people used to the old rules. Some parts of the old rules had textual problems, such as errors in numbering or phrasing, that ostensibly required correction. For the sake of avoiding disruptions for the audience, however, the restyling team let these problems stay. While not called "sacred," these familiar items help readers already familiar with the Evidence Rules to continue to use them efficiently. Capra described a textual problem involving Rule 803, which describes exceptions to the rule against hearsay:

> So Joe's [Kimble] right, you don't follow a number with a number. 803(1) should be 803(a). But what's the value of changing that in this circumstance? . . . So now it's 803(a) instead of 803(1), excited utterance [Rule 803(2)] would be 803(b), and so forth. All of those options are bad options for the obvious reason that this is a very often-used rule, and if you change the numbering in any way, what you've done is you've disrupted electronic searches. You disrupted the expectations of all the parties that are applying this on a day-to-day basis. (Douglas et al. 2012, 1452)

Capra's example shows how committee members left the numbering errors in Rule 803 uncorrected in order to allow users of the Evidence Rules to find and return to that section's contents more easily. The committee chose not to disrupt readers' expectations by fixing the errors. Judge Marilyn L. Huff said, "The Rule

803 example illustrates the practical point that the restyling should help, not confuse, the reader" (Douglas et al. 2012, 1464).

In some cases, the team allowed words or phrases to remain even when they were not especially plain or clear. As Judge Ericksen said, "There are times in the Rules when the passive voice really is the best voice, and when a lack of precision is precisely what is required. We often argued and struggled over words and phrases in the Rules that have acquired subtle meanings to courtroom lawyers and judges that are not found in a dictionary" (Douglas et al. 2012, 1457–58). The committee exercised discretion in deciding when to prefer a passive verb or an imprecise word, and this discretion helped the committee meet the audience's expectations for the content and meaning of certain parts of the Rules.

Another textual problem involves ambiguity and vagueness. One of the general goals of restyling, and of plain-language projects in general, is to remove ambiguity and vagueness (Garner 2007, 46). An example of an ambiguous word is "shall," which appears frequently in legal documents. Legal professionals see several meanings in "shall," including must do something, may do something, and will do something (Garner 2007, 29–30). In the symposium on the Evidence Rules, Professor Edward H. Cooper of the University of Michigan, reporter for the Advisory Committee on Civil Rules, gave an example from the Civil Rules. Restyling participants changed an instance of "shall" in Rule 56 to "should." After the restyled rules took effect, the Advisory Committee changed it back to "shall" during the process to make substantive rule changes. The Advisory Committee found reason to prefer the flexibility—some would say ambiguity—of "shall" (Douglas et al. 2012, 1479–80). Kimble (2013) used this same example to show that plain-language advocates will not win every battle for plainness that they fight, so to speak; plain-language advocates must choose their battles and try to keep larger project goals in mind. Yet Kimble pointed out that this occurrence of "shall" is the only one that remains in the four restyled sets of rules (pers. comm., December 2, 2013). This example from the Civil Rules shows that while plain-language advocates will not get every textual change they want, they are likely to get many of them by following effective processes.

Seeking and Considering Feedback

Many plain-language experts routinely seek feedback on the documents they create, but it is especially important to seek feedback on documents that affect many people. Each restyling process step involved collecting feedback on proposed changes. As the restyling consultant to the Style Subcommittee, Kimble received volumes of feedback from committee members. The public comment period for the restyled rules also generated feedback from interested parties, such as judges, lawyers, and law professors. Over four years, Kimble said, hundreds of documents entered the archive for the restyling project (pers. comm., October 23, 2013.) Kimble and his colleagues carefully considered each amount of feedback they received. Said Kimble, "I spent a lot of time—a lot of time—responding to comments. One memo and email after another. And that was all to the good" (pers. comm., October 23, 2013). When done with sincerity, the process of seeking and considering feedback forces any editor to continually reconsider the value and the validity of the reasons behind changes made to a text.

Using Style Guides

Another standard process for people who regularly work on plain-language projects is to use style guides, but it is worth noting. Judge Hinkle ensured that team members followed Garner's "Guidelines for Drafting and Editing Court Rules" (2007). These guidelines give general advice about writing in plain language in addition to specific advice about creating appropriate structures in the rules. Kimble (2013) mentioned that Hinkle's decision helped improve the process and the product of the restyling effort. Hinkle said that it helped to have the principles written in advance, even though Kimble knew them well. Hinkle added that using Garner's guide helped keep the approach to the Evidence Rules consistent with the earlier restyling projects (pers. comm., November 28, 2013). One reason that organizations adopt style guides is to increase consistency among the documents written by many different authors. Kimble specifically mentioned that inconsistencies plagued the old Evidence Rules (2009), so Hinkle's adoption of Garner's (2007) guidelines was especially appropriate.

Organizational Culture and Plain Language in the Restyling Project

The culture of the Advisory Committee on Evidence Rules helped ensure that the group restyled the Evidence Rules successfully. The successes of other advisory committees on their rules provided helpful precedents. Hinkle described the colleagues involved with the Evidence Rules project:

> We had a group of high achievers with a "can do" attitude. We had seen the success of the earlier restyling projects and knew those restyled rules were better. And we looked at samples of several proposed restyled evidence rules and saw that they were much improved. One can argue about the cost of change. But it is hard to argue that the new rules are not better than the old ones. (pers. comm., November 28, 2013)

The task of restyling the Evidence Rules must have been daunting at times, especially as it took four years to complete. The culture among the committee members helped them persevere.

The Standing Committee's first successful restyling project led to others. The Standing Committee first attempted to restyle the Civil Rules in the early

1990s, but the project stalled. The Standing Committee then shifted attention to the Appellate Rules, which Kimble said are probably less sensitive and certainly shorter than the Civil Rules. After that project was successful, Kimble said, the Criminal Rules, Civil Rules, and Evidence Rules followed (pers. comm., October 23, 2013). The success of restyling the Appellate Rules created momentum that benefitted the other restyling projects.

In the symposium on the restyled Evidence Rules, some participants stated that they had not always supported the restyling efforts. Judge Reena A. Raggi joined the project after the restyling project had begun. She said she openly expressed concerns about the high "transaction costs" of inconveniencing people who were used to the existing rules. She disagreed with Kimble's assessment that the old rules were poorly drafted and embarrassing. Raggi also stated that in some cases, she did not think that restyling improved the rules noticeably (Douglas et al. 2012, 1465–70). Judge Harris L. Hartz said that a restyled rule may receive new and unexpected interpretations. He provided examples of how restyling may create an ambiguity, eliminate an ambiguity, or inadvertently change the rule (Douglas et al. 2012, 1483–86). In one instance, Hartz showed how a restyled passage in the Rules of Civil Procedure changed the definition of how a pro se defendant should serve a lawsuit notice to the US Attorney General. Judge Fitzwater pointed out that Raggi and Hartz contributed substantially to the project in spite of their reservations. Fitzwater said Raggi, in particular, caught some problems that others had overlooked: "Everybody worked hard and did it in complete good faith with no hidden agenda. And I think it worked out" (Douglas et al. 2012, 1447). The narrow scope of style changes and the clear lines of authority for the restyling project likely helped Raggi and Hartz to contribute in spite of their reservations.

Kimble pointed out that the culture of the legal community extends to teachers and students at law schools (pers. comm., December 1, 2013). Two professors at the symposium mentioned that restyled rules are good for students. Professor W. Jeremy Counseller of Baylor Law School had taught his students with both the old and the restyled rules of Civil Procedure. He asked students which version of the rules they preferred, and students unanimously supported the restyled rules (Douglas et al. 2012, 1509). Professor Roger C. Park of the University of California's Hastings College of the Law provided data about student performance on two test questions about evidence rules (Douglas et al. 2012, 1545–47). Using a chi-square test, Park found that students taught with the restyled rules performed significantly better than did students taught with the old rules on one question, but noticeably worse on a second question (without reaching statistical significance). In discussing these results, Park wondered whether the committee had foregone changes to Rule 804(b)(1), the subject of the second test question, in order to avoid excessive transaction costs, as Judge Raggi termed them (Douglas et al. 2012, 1496). Unfortunately, the symposium transcript does not answer Park's question.

Judge Hinkle said it helped to take a long-term view during the process of restyling the Evidence Rules. "The arguments against restyling were all transitory: the project would consume resources, and lawyers and judges who knew the rules would have to learn the new provisions. The arguments for restyling were far more substantial: the new rules would be better. The transitory costs are mostly gone. The benefits will be here for years to come" (pers. comm., November 28, 2013). The long-term view helped committee members overcome the difficulties they faced in the short term.

Linking Ethics and Plain Language in the Restyling Project

The participants in the restyling process acted ethically by upholding expectations for professional ethical conduct and by conducting important work in ethical ways. Judge Hinkle commented on ethical conduct: "The restyling project involved ethics in the same way that handling cases as a district judge involves ethics: the participants needed intellectual honesty and a good work ethic" (pers. comm., November 28, 2013). Hinkle links professional behavior, focused on accomplishing the task at hand, with ethical behavior. Kimble said participants followed ethical processes:

> First, the process was completely open. It involved a range of interested parties: judges, lawyers, law professors, and legal drafting experts. All the minutes and various drafts were available online. All the rules were available for public comment. And the Advisory Committee convened a symposium that included both supporters and opponents. At any rate, the process took four years and produced hundreds of documents in the archive. So what does all that suggest? A commitment to openness and getting the best result. (pers. comm., October 23, 2013)

Kimble's comments show that the committee worked in a way that supported professional integrity and kept their actions above reproach.

The committees did have to vote occasionally on whether a particular change was substantive or purely stylistic. Kimble said that the votes were not always unanimous. In the absence of unanimity, the majority carefully considered the concerns of the minority and avoided changes that left a substantial minority ill at ease (Kimble 2013). This approach to voting and discussion shows that the committee truly valued the dialogue among its members. Rather than simply acting on the basis of numbers of votes, the committee members recognized the value in each other's perspectives.

Kimble said that he and other participants in restyling drew on their experiences to make the Evidence Rules as clear and useful as possible. All of them realized that the project significantly affected their colleagues, clients, and students, as well as the public at large. Each participant brought many years of experience to the project. Reflecting on the four-year project, Kimble said the process shows that the restyling participants tried to restyle for the audience, not at the audience. Kimble believes this restyling follows Buber's I–You mode of communication rather than the I–It mode (pers. comm., October 23, 2013). The committee did not simply restyle the Evidence Rules and foist them upon the public, as a supervisor might speak at a subordinate in the I–It mode. The committee instead sought dialogue through public comment, and it considered each comment received. The committee respected its varied audience of legal practitioners, scholars, and students.

Conclusion: Key Takeaways about Plain Language from the Project to Restyle the Federal Rules of Evidence

The Committee on Rules of Practice and Procedure (the Standing Committee) within the Judicial Conference does important work to serve the US federal court system. Members of the Standing Committee and subcommittees come from a range of areas within the legal profession. Although few members have training or experience in plain-language writing, they all share the goal of ensuring that their assigned rules and procedures support the goals of fairness and justice. The ethical principles applied in the restyling project include respecting values of fairness and justice, respecting the rights of all individuals affected by the Evidence Rules, and demonstrating ethical professional behavior to collaborators. Here are four main ideas that another organization can take from this project to restyle the Evidence Rules:

1. **Use prior projects as evidence that plain language works.** The restyling work on the Appellate Rules, Criminal Rules, and Civil Rules provided precedents for the work on the Evidence Rules. Each successful restyling project helped create momentum for the next one. Even when a plain-language document comes from a different organization, it can provide evidence that plain-language projects are worth investments of time and effort.

2. **Learn when to argue for plainer language and when to accept the status quo; consider the costs of change.** Kimble (2013) made this point at the PLAIN 2013 conference. Plain-language professionals are unlikely to win approval for every change they seek. Thus, focus on larger goals of improving a document. When revising an existing document, acknowledge that changes may make it more difficult for readers who are familiar with the old version. Ensure that the transaction costs of change will not be too high for too long.

3. **Whenever possible, use an open revision process with clear procedures and guidelines.** The Advisory Committee on Evidence Rules used a clear process with clear lines of responsibility for decision making. The process allowed all participants to voice their concerns, and the procedures helped

project members to separate matters of style from matters of substance. The Advisory Committee avoided making changes in response to vague suggestions that someone simply did not like the wording. The clear process also helped to sustain the work over four years. By keeping records of discussions and inviting public comment, restyling participants showed respect to each other and showed concern for the success of the final product.

4. **Hire or appoint a plain-language expert to prepare the first draft**. A plain-language expert will improve the content by exposing ambiguities, inconsistencies, and other uncertainties. Using an expert will make the content much more clear and readable. Bringing a plain-language expert in at the end of a big writing project suggests that plain language is merely cosmetic. Clear procedures and guidelines are important to the expert's work as well. Consider all views and objections, but unless the expert's draft changes or misstates the meaning, it should prevail.

The legal system in the US is vast and complex. State and federal court systems are bureaucracies that affect citizens' lives substantially. The language of the legal system goes back for centuries in some cases, and its jargon is unfamiliar to most citizens—even though citizens often feel the effects of laws and statutes more than do the attorneys, judges, clerks, and others who work within the legal sector. Because they lack familiarity with the system and its language, citizens frequently turn to professionals when a legal issue arises. Laws and court decisions can significantly affect individuals' freedoms and their opportunities to exercise the rights of citizenship. Critical court decisions can determine, for example, whether one owes or will receive sums of money, whether one remains free or goes to prison, or whether a company will be able to manufacture and sell a product.

The restyled Evidence Rules affect the US legal bureaucracy from the inside. As more law students, attorneys, and judges become familiar with the restyled Evidence Rules, these individuals will be able to understand, interpret, and follow the rules. Through adhering to a structured process, using pertinent plain-language resources, fostering dialogue between committee members and also with the public, carefully considering feedback, and carefully managing the scope of changes, the Style Subcommittee of the Standing Committee on Rules of Practice and Procedure created a plain-language document that will benefit participants in federal court trials for years to come.

Questions and Exercises

1. Does your organization have a significant document that you want to revise? Will the process for revising this document be lengthy and challenging? Look around for a similar document revised into plain language. Maybe you will find one in your organization; maybe you can find one at another

organization. Look for winners of plain-language awards such as the Clear-Mark awards from the Center for Plain Language. After you have found a similar plain-language document, write a plan of 300 to 500 words that describes how you will garner support for the revision process and how you will design it.

2. Look for some policies and procedures that affect you, whether at work or in an organization to which you belong. This might be a set of bylaws or a policy on completing certain kinds of work. Analyze this document. Identify key features of it, such as its primary and secondary audiences, its purpose, and its navigation tools and design; the frequency with which audiences use it; and the effects that it has on its audiences. After you analyze it, draft a plan of 400 to 600 words to either restyle or revise the document. Your plan should include a list of people to participate in the project, a set of objectives for the project, an estimate of the time required for the project, and a list of any organizational political conditions that affect the plan. Get feedback on this plan from someone you trust, and then consider sharing it with someone who can help you gain approval for the project.

7

PROFILE–COMMONTERMS

CommonTerms is a volunteer-led project based in Sweden that addresses the complex, convoluted terms and conditions to which consumers must agree before using software and some online services. CommonTerms is led by Pär Lannerö, who is also a senior partner at the Swedish firm Metamatrix, a consultancy focused on web accessibility and user experience.

Terms and conditions are legally binding documents. Most providers of online services write these documents in voluminous legalese; such documents serve the providers' interests but frustrate the end users. CommonTerms addresses one aspect of the problems related to online terms and conditions. The Common-Terms online tool generates a one-screen preview of a company's "terms of service" in plain English before giving users the option to view the complete terms and conditions. This tool generates HTML code that a company may add to its website to provide the preview to its users. CommonTerms has released the tool in its beta version, and the group is working toward a release of version 1.0.

This profile of CommonTerms shows how the group combines foundation-supported work with volunteer efforts in an interesting way as it addresses a pervasive and important problem. CommonTerms also uses the internet to promote its cause, to recruit volunteers and supporters, to gather feedback on the plain-language tool it is developing, and to share that tool with web developers. Some work on the CommonTerms project occurs in face-to-face sessions while other work takes place asynchronously online and off-line. On its website, CommonTerms (2013c) notes that it has been inspired by Creative Commons, which provides ways for content creators to indicate the degree to which others may use and even transform their work. CommonTerms also notes that it cooperates with many other groups that do work related to documents concerning online terms and conditions. Although CommonTerms has not released the source code for its preview generator as a form of open-source software development, the HTML code it does provide fits into the collaborative, cooperative spirit of open-source programming and Creative Commons content sharing.

This profile shows how CommonTerms creates plain-language summaries of terms and conditions, allowing users to make better-informed decisions in a timely manner about the information they share online. It describes how the content produced by CommonTerms addresses specific BUROC situations, how the group provides its content, and how organizational culture and ethics affect the work of CommonTerms. It closes with lessons that plain-language professionals can take away from this profile.

Background of CommonTerms

Pär Lannerö has worked as a programmer and a computer specialist for decades. He studied computer science in the early 1990s, and he soon started using the internet. "When I first saw the Mosaic web browser in 1993," Lannerö said, "I understood the web was going to be important" (pers. comm., November 5, 2013). Lannerö learned all he could about the web and the internet. He started teaching others about using online resources. Sweden's national education agency, Statens Skolverk, hired him to help teachers and students learn about the web and the internet. Lannerö also cowrote an introductory book for Statens Skolverk about using the internet (Ericsson and Lannerö 1997) that educational agencies later translated into English and Norwegian.

In 1999, Lannerö started Metamatrix with four other internet experts he had met through his work in education. The firm initially focused on internet calendaring systems. Over time, Metamatrix moved from creating structured information and XML systems to developing web-based services. Metamatrix continues to work on the web; the company currently focuses on design for user experience and online accessibility for individuals with disabilities.

Lannerö said the CommonTerms project grew from a session about sustainability on the web that he led at the Sweden Social Web Camp (SSWC) in 2010 (pers. comm., November 5, 2013). SSWC is an informal conference, or unconference, about social media and internet technologies. On the bus to SSWC, Lannerö sat next to Tomas Bjelkeman, an environmentalist and the manager of a foundation called Akvo that integrates internet technologies with foreign aid. Their discussion—on the bus and at SSWC—about nonsustainable online practices included the difficulties of password management and problems of unwieldy complex terms and conditions. Lannerö defined online sustainability as the responsible use and management of online resources, including everything from programming code and user passwords to informational content and effective user-interface design (pers. comm., November 14, 2013). After SSWC, Lannerö wrote a blog post that outlined the framework for a potential solution that would be more sustainable for companies and their customers—a proposal for common terms of service that all parties can easily understand (Lannerö 2011). After receiving encouraging responses to the SSWC session and the blog post, Lannerö sought and received funding from Sweden's Internet Infrastructure Foundation (the IIS) to develop and refine his proposed ideas through the CommonTerms project.

CommonTerms raises awareness about the problems of voluminous, legalese terms and conditions through a companion project called the Biggest Lie. At the Biggest Lie's website, http://www.biggestlie.com, people have the opportunity to confess to the "biggest lie" on the web: they have clicked to indicate that they have read and agreed to terms and conditions without actually reading them. The Biggest Lie site uses CommonTerms's descriptions of the problems with inaccessible terms and conditions, and it lists potential solutions to those problems from CommonTerms and other groups.

CommonTerms addresses a situation that faces virtually every consumer who uses software on internet-connected devices, from laptops and desktop computers to tablets and smartphones. Terms and conditions apply to software (including the many applications, or apps, downloaded onto smartphones and tablet computers), social web platforms such as Facebook and Twitter, and websites. These documents tend to be long, dense, and written in the language of the legal bureaucracy; many terms are unfamiliar to consumers. For example, the following sentence (appearing in all capital letters) from the terms of use for Instagram (2013), the popular internet photo-sharing application, contains terms such as "release" and "claims" that are likely to be unfamiliar to readers, along with the quoted material from the Civil Code of California:

BY ACCESSING THE SERVICE, YOU UNDERSTAND THAT YOU MAY BE WAIVING RIGHTS WITH RESPECT TO CLAIMS THAT ARE AT THIS TIME UNKNOWN OR UNSUSPECTED, AND IN ACCORDANCE WITH SUCH WAIVER, YOU ACKNOWLEDGE THAT YOU HAVE READ AND UNDERSTAND, AND HEREBY EXPRESSLY WAIVE, THE BENEFITS OF SECTION 1542 OF THE CIVIL CODE OF CALIFORNIA, AND ANY SIMILAR LAW OF ANY STATE OR TERRITORY, WHICH PROVIDES AS FOLLOWS: "A GENERAL RELEASE DOES NOT EXTEND TO CLAIMS WHICH THE CREDITOR DOES NOT KNOW OR SUSPECT TO EXIST IN HIS FAVOR AT THE TIME OF EXECUTING THE RELEASE, WHICH IF KNOWN BY HIM MUST HAVE MATERIALLY AFFECTED HIS SETTLEMENT WITH THE DEBTOR."

Public policy scholars Aleecia M. McDonald and Lorrie Faith Cranor estimated that American internet users would need 201 hours each year to read privacy policies word for word each time they visited a new website. The value of this time for an estimated 221 million US internet users is $781 billion. McDonald and Cranor (2008) wrote that in 2007, online sales and advertising in the US totaled

$260 billion; thus, "the current policy decisions surrounding online privacy suggest that internet users should give up an estimated $781 billion of their time to protect themselves from an industry worth substantially less" (563). To use these online technologies, consumers divulge varying levels of their personal information, information that they have rights to control. If terms of service change, or if they want to use a new service or a new online tool, consumers face critical choices that they must make quickly. The following list from the CommonTerms (2013c) website summarizes the major problems with terms and conditions:

- Users get exploited: Most agreements have been formulated only by providers, and therefore [are] **not protecting the interest of users.** Even if terms are fair, some users will regret having signed the agreement because they were expecting something else.
- Users get excluded: Some users who do not want to accept an agreement they cannot understand or do not have the time to read will **stay away from great services,** thus reducing market size and contributing to the digital divide.
- Limiting value: Some users, not knowing the terms, will be restrictive in their use, and only get **limited benefit** from the services. How fun would a social network be if you didn't trust it with some personal data?
- Wasting time: Some users spend a lot of time actually reading the agreements, and service providers spend time writing for almost nobody.
- Lessening competition: Websites and other online services should be able to compete with great terms and agreements, but today that's hard because nobody has a clue what they agree to.
- Eroding respect for contracts: Some users may **become less respectful of agreements** in general, having ignored so many of them. (emphasis in the original)

Consumers are making stronger complaints against terms and conditions they do not understand. In a description of the beta version of CommonTerms's preview generator, Lannerö (2013) wrote that the Swedish agency for consumers has received complaints from consumers who did not understand conditions to which they had agreed. When online service providers such as Google, Facebook, and Twitter change the terms of use for their services, users sometimes lead grassroots movements against the changes, government agencies may investigate, and lawyers may even seek class-action lawsuits against the company changing the terms.

Personnel Who Create Plain-Language Content for CommonTerms

CommonTerms combines funded work and volunteer efforts. In addition to Lannerö leading and actively participating in the project, key personnel include a project administrator to manage invoices and reports and an interaction designer to manage the online experience of using CommonTerms tools. Lannerö said that some project members have volunteered after reading about CommonTerms online or hearing a conference presentation while others joined after referrals from the IIS (pers. comm., November 5, 2013). The CommonTerms (2013a) website identifies almost 40 people who have participated in the project. They include usability experts, attorneys, law professors and law students, Creative Commons copyright advocates, graphic designers, online privacy advocates, and leaders of related online privacy projects. Key activities include programming, icon design, content development, and interaction design.

Practices and Processes for Creating Plain-Language Content for CommonTerms

The content development practices for CommonTerms include focusing on items that the audience needs most, delivering content in appropriate ways, ensuring that visuals support the content and help users, observing users to learn about the content that they want and how they will use it, learning from others and sharing information with them, and continuing to improve the content.

Focusing on Most-Needed Items

CommonTerms focuses on terms that users are likely to see in an online terms-of-service document. Team members started by identifying a set of these documents to analyze. As Lannerö (2012) describes it, they sought a representative selection of websites. They started with statistics that identify the activities people do most on the internet. Then they used Google AdPlanner to identify websites popular for those activities. Finally, they complemented that list of websites with a few sites from other important categories. For example, they identified some lesser known local websites because people use such sites quite often. Another category of sites they included was startup companies (12).

The team then analyzed each site's terms-of-service documents. They developed an SQL (structured query language) database with a web-based interface. The software allowed team members to analyze key terms in the database, identify the topic each key term addresses, and link recurring topics in contracts (e.g., age limits) with various specifications that a user must meet (e.g., age 18 or older) in order to fulfill the contract (Lannerö 2012, 13–14). The software also allowed team members to work online remotely and asynchronously.

With a database of 22 terms-of-service documents, the team found 450 key terms repeated throughout. Although a larger group of websites might have yielded an even larger number of key terms, the CommonTerms team thought that 450 key terms provided a good foundation for future work (Lannerö 2012, 16). Between the team's alpha proposal and its beta version, the version that

FIGURE 7.1 Sample preview of a terms-of-service document generated by the beta version of the CommonTerms online tool.

included the content generator, the team decided to focus on seven sections of typical terms-of-service documents:

1. Business model
2. Payment and cancellation
3. Agreement details
4. Content and copyright
5. Privacy and security
6. User restrictions
7. Certificates and guarantees (Lannerö 2013, 22)

At the same time, the online generator allows companies to add their own terms to their summaries (Lannerö 2013, 25–28). Figure 7.1 shows a sample preview of a terms-of-service document.

Delivering Content Appropriately

Sustainability is one of Lannerö's overriding concerns with the ways that people use the internet and the tools and the ways that programmers create internet tools. To overwhelm consumers with information in terms-of-service documents is not sustainable; overwhelmed consumers either lie about reading the terms or decline the use of a service or product. Similarly, companies cannot sustainably invest considerable time in writing, updating, and presenting their online service agreements. The online generator provides a sustainable method for creating preview terms-of-service documents for users to review (Lannerö 2013, 27). It combines the efficiency of online forms with the guidance of a software wizard

FIGURE 7.2 The CommonTerms online tool guides users through creating a preview of their terms and conditions.

interface. Figure 7.2 shows a partial list of the common terms from which users may select to build a preview to their terms and conditions.

A company may take the HTML code from the generator and insert it on its website. The code links back to a CommonTerms server, but the preview appears on the company website for users to see. Lannerö said that it made sense to create HTML code for others to use quickly and easily. As examples, he pointed to online services such as YouTube and SlideShare that provide quick access to HTML code that others can easily include on their own websites (pers. comm., November 14, 2013). By providing easy ways to share and imbed their content, YouTube and SlideShare engage users and attract more viewers to their content. While programmers can provide online content in a number of formats and file types, CommonTerms chose to deliver its content in HTML code because its web developers can incorporate that code quickly and easily.

Creating Useful Visuals

From the beginning, the CommonTerms team has sought to use icons to convey information in addition to the verbal descriptions of key terms. The web provides several examples of icons created to convey information about personal data use, online privacy, data collection, and data sharing. The CommonTerms team took inspiration from these projects (Lannerö 2013, 8–10) and collaborated with some other designers to create icons. In the beginning, Lannerö expected to have icons for every key contract term, but the team soon realized that the terms are too numerous: "Initially the idea was to use icons for every single common term pattern [found repeatedly in the sample of terms-of-service documents]. We gave up on this when we realized contracts contain so many different terms. Having thousands of icons simply wouldn't be meaningful. Now we use icons only to support visual navigation between term categories" (pers. comm., November 5, 2013). These icons help readers to navigate through and to understand the content of the preview by complementing the verbal description of each section heading.

Testing the Content

CommonTerms conducted two types of observations to understand how users might react to a preview of terms-of-service documentation. The first observation included testing a prototype website with 10 users. Researchers Hanna Arkestål and Carl Törnquist from Stockholm University presented to users a mock website with a CommonTerms preview button. They asked users to sign up for an account on the mock website and to think aloud while they worked. If users did not spontaneously click the preview button, researchers asked them to do so afterward. Researchers recorded the users' reactions to the preview button and the one-screen summary (Lannerö 2013, 18).

Arkestål and Törnquist found that only one person in their small sample used the preview without a prompt to do so. After users learned about the preview and understood its purpose, they responded positively; 6 of 10 users said that they would use it. Arkestål and Törnquist thought their users had become accustomed to skipping license agreements, but they hoped that a preview of the agreement terms could help break this habit. Arkestål and Törnquist (2012) reasoned that a preview-supplied understanding would lead to at least a partial understanding of the full agreement. They also recommended presenting a panel with icons that identify the areas of the preview that matter to users most instead of a button simply labeled "Preview Terms" (Lannerö 2013, 18–19).

The second observation involved an alpha test with an online knowledge-management service called Refinder, produced by the Austrian company Gnowsis. Gnowsis placed a CommonTerms preview button on the screen used to sign up for a Refinder account. Log files for Refinder indicate that approximately 10% of users clicked the preview button. While not a large percentage of users, this number is noticeably higher than the 1% to 2% of users who normally open the terms-of-service or privacy-policy documents (Lannerö 2013, 19). Through the Refinder test with Gnowsis, the CommonTerms team learned valuable lessons about how a company might want to incorporate a preview into its website. The team also learned about balancing the preferences of companies and users. Companies will not want the preview button to be too prominent because they want users to focus on signing up for online accounts to use their services. At the same time, some users appreciate the preview and see it as a signal of a company's trustworthiness. In a sense, CommonTerms serves two groups of users at once: those who create websites and web services and those who use them. Both groups must elect to use CommonTerms; CommonTerms cannot compel either group. Through testing, CommonTerms learned about the practical concerns of the website owners and the consumers (20).

Learning from and Sharing with Others

The problems related to terms-of-service documents, privacy policies, end-user license agreements (EULAs) for software purchases, and other similar documents online are interrelated and interconnected. The companies creating the online services and their clients, the users, are likely to be at odds over many issues. The CommonTerms team is one of the many groups trying to improve the situation in which consumers agree to terms that they do not read or understand well. Some other groups focus on creating icons and graphics to help consumers understand the policies; others have worked to make the text in these policies plainer and more accessible. Still other groups have worked to raise awareness about the problems and to seek changes in policy. Many such groups appear in lists on the CommonTerms (2013d) website and in the beta proposal for CommonTerms (Lannerö 2013).

CommonTerms joined with other groups and individuals to form Open Notice, a forum for collaborating and sharing ideas about legal documents that regulate online activities. Open Notice (2013) is "a community effort calling for an open, global, and public infrastructure for legally required notices." The group has regular conference calls, an email list, and occasional in-person meetings for work on particular projects and issues. While the CommonTerms team members do seek to create a successful product in its preview generator, they are more interested in helping to solve problems related to confusing, convoluted online legal documents than to advocate for their specific solutions to a problem.

Continuously Improving the Content

Like other online tools and services, the CommonTerms preview generator has not reached stasis. As a tool in its beta version, the preview generator will continue to change. The CommonTerms team posted a list of tasks it will complete in the future (CommonTerms 2013b). One avenue the team will explore is standardization of online legal documents through W3C, the World Wide Web Consortium. The team will continue to seek feedback from consumers, website owners, and fellow consumer advocates on the preview generator and the database of common terms. Lannerö said that having expert writers and legal experts review and improve the database will be particularly important (pers. comm., November 5, 2013). The team may also translate the database into languages other than English. Lannerö (2013) wants to ensure that the terms database will be culturally appropriate for audiences: "Formulations of standard terms will need to be checked and rechecked by people from several different cultures in order to be correctly understood by as many as possible" (34). Additionally, the team will promote the cause of CommonTerms and will seek renewed funding to support the project.

As the CommonTerms team strives to improve its work, the team faces at least five main challenges (Lannerö 2013, 34–35). First is the challenge inherent to communicating legal information: it is often difficult to summarize legal information. Some believe that it can be risky to provide simplified versions of legal documents. It is likewise difficult to convey that information in culturally appropriate ways and in more than one language. Collaboration is a second challenge faced by CommonTerms and the groups in Open Notice. Geographic distance and cultural differences can make it challenging to work cooperatively. Additionally, online legal documents affect many professional disciplines, and people from many disciplines will need to contribute in order for CommonTerms and other projects to succeed. Inertia and possible opposition can create a third challenge. Established companies and law firms are likely to support the current status quo of online legal documents; they might actively oppose change. A fourth challenge is a lack of visibility for the issue. Although problems arise when consumers fail to read online agreements, other problems such as piracy and illegal file sharing often appear more important. The fifth challenge, closely related to the fourth, involves finding funds and energy to sustain CommonTerms. Many similar projects have stopped, likely for lack of resources.

Organizational Culture and Plain Language for CommonTerms

In a way, the unconference presentation about online sustainability that Lannerö gave in 2010 established the culture embodied in the CommonTerms project. CommonTerms seeks to provide an efficient, sustainable solution to several interrelated problems that defy sustainability. CommonTerms's innovative generator provides a preview of terms of service in plain language. Plain language is an essential part of the sustainable solution that CommonTerms continues to develop and refine.

In an interview with a UK consultancy that focuses on the changing practices for managing users' personal data, Lannerö explained part of his motivation for pursuing sustainable practices for online legal agreements. He said, "I think highly of the ideas put forward by Doc Searls and the VRM [vendor relationship management] movement: users should not be forced to take or leave whatever vendors serve. Instead they should be on equal footing and be able to negotiate on those terms" (Ctrl-Shift 2013). Lannerö and CommonTerms advocate directly for users' rights.

Although CommonTerms and other members of the Open Notice group primarily advance the interests of consumers, these groups must also appeal to the businesses and organizations providing online services to consumers. These groups must demonstrate that sustainable legal-documentation practices simultaneously benefit businesses and consumers. Plain language practices support clear communication, and clear communication will help businesses and consumers to sustainably conduct activities online.

Although the law governs contracts, contracts also provide an opportunity for the agreeing parties to act ethically. Effective contracts record a meeting of the minds of all parties involved. Lannerö said that unfortunately the parties who do not write the contracts will not know all the details and their consequences. He said an ethical company will make sure the terms are fair and will communicate clearly anything of substance that the other party could not reasonably expect in the contract (pers. comm., November 5, 2013).

In Lannerö's view, users have less of an ethical obligation than do companies providing services and offering terms of use. Nevertheless, "if you assert having read and agreed to a contract (possibly thus accepting responsibilities towards the provider) then you are giving yourself an ethical obligation to fulfill these responsibilities. This may be difficult to do if you haven't actually read the contract. And in most moral systems, I guess, lying in a contracting situation is not ethically sound" (pers. comm., November 5, 2013). Lannerö appears to hold the companies to a higher ethical standard because they have more power than the consumers do. He has stated that consumers should have equal footing with the companies (Ctrl-Shift 2013). Although consumers lack the companies' advantageous positions, Lannerö does not absolve them of their ethical responsibilities.

The preview generator from CommonTerms provides a way to help companies and consumers have a dialogue about terms of service, privacy policies, and the like. When companies and their consumers understand each other, they are more likely to respect each other and have a sustainable business relationship.

Conclusion: Key Takeaways about Plain Language from CommonTerms

CommonTerms is an ad hoc organization supported by foundation donations and volunteer support. CommonTerms personnel share information with other like-minded groups through the Open Notice forum, social media, and other means. CommonTerms shows how techniques of plain language and clear communication apply to important relationships between companies and consumers. The CommonTerms project involves ethical principles of respecting the rights of consumers, addressing the power differential between online service providers and consumers, communicating clearly to avoid misleading consumers, and promoting the honesty of service providers and consumers. I have identified three lessons of the CommonTerms project that other organizations may apply:

1. **If you are dealing with a complex problem, focus on one part of that problem.** Many people and organizations around the world have identified problems associated with the terms and conditions consumers must

accept to use software and online services. Providers of software and services want to protect their interests, and they may have little incentive to change the complex, convoluted documents they use. Increasingly, however, consumers want to understand their rights better and to exercise their rights more carefully. The CommonTerms project focuses on one aspect of this provider–consumer divide—a preview summary of terms and conditions—rather than attempts to address the entire problem. The CommonTerms team's successes with the preview generator have given them knowledge and experience to share with other like-minded individuals on similar projects.

2.   **Deliver content in a usable format.** Consumers encounter a lot of terms-of-service documentation online. The CommonTerms preview generator creates content that is instantly available for use online. By offering content in this format, CommonTerms increases the likelihood that others will use their content.

3.   **Make plain language a part of your organization's plan to improve.** The CommonTerms project has moved from the alpha stage to the beta stage; eventually, it will reach version 1.0. One part of moving past the beta stage involves improving the clarity, accuracy, and cultural relevance of the CommonTerms database. While it would be natural to focus on improving the coding and the technical performance of the CommonTerms preview generator, Lannerö realizes that development of the content must keep pace with the online tool itself.

In many ways, the internet is an environment that promotes equal access and individual freedom. Once online, individuals may seek information, join virtual groups, learn new skills, find entertainment—the possibilities are almost endless. As more companies have gone online to create services and attract customers, and as more of them want to collect data from online users, they have developed policies and disclosures about those activities. The vast majority of those documents reflect companies' desires to protect themselves from legal action, and they use the bureaucratic language of the legal system instead of consumers' typical language. Consumers, who are unfamiliar with legal jargon and with legal documents in general, frequently tell what Lannerö and his colleagues call "The Biggest Lie" by asserting that they have read documents without reading them. These critical consumer decisions affect both the quantity and the quality of the information they reveal. Consumers have rights to choose which online services they will use, and online service providers should not coerce them by forcing choices between reading unfamiliar, unwieldy legal documents on one hand and lying about having read them on the other.

The CommonTerms project tries to bridge the gap between companies and consumers. Contracts are supposed to represent agreements between parties, and the CommonTerms preview generator puts the major concepts of the terms and conditions into consumers' language. The preview generator supports dialogue and shared understanding. The terms-and-conditions preview generator helps companies satisfy their desires to protect their interests while giving consumers a better sense of the terms to which they agree. It is ethical for companies to help their consumers understand terms and conditions, just as it is ethical for consumers to be honest about the agreements they make.

Questions and Exercises

1. The CommonTerms team reached out to members of related projects and created the Open Notice forum. Open Notice provides members opportunities to learn and share information. Are you an individual member of any professional associations? Does your organization have relationships with similar organizations? Identify three ways in which you or your company can improve your networking, and then identify the first steps you can take toward improvements. Record this information in a memo of around 300 words.

2. The CommonTerms team had the technical expertise to create its own tool to analyze terms-of-service documents from several websites. Do you have tools to support the work and analysis you need to do? If you lack tools or need better ones, use websites to search for appropriate tools. Compare available tools using criteria that are meaningful and specific to your situation. After you find some tools, think about how you might obtain them, and then draft an email or memo of 200 to 300 words to your boss requesting resources to obtain them.

8

PROFILE—HEALTH LITERACY MISSOURI

Health Literacy Missouri (HLM) is a nonprofit organization based in St. Louis. Working with medical providers and other organizations across the state, HLM's goal is to help Missourians make better health decisions every day. Whereas Healthwise, discussed in chapter 4, strives to help people make better health decisions by providing clear and informative content, HLM's approach focuses on training, reviewing clients' documents, and raising awareness about health literacy. HLM helps people to improve their own health literacy, and it trains medical providers and other medical companies to communicate effectively with audiences who often lack strong health-literacy skills.

I studied HLM's plain-language activities because those activities involve more than providing content to clients. HLM integrates plain language and health literacy in innovative ways that empower clients to improve their own communication skills. The organization enacts a dialogic, ethical approach to communication by training clients to communicate effectively with people who have low health literacy. Like other organizations working in plain language and health literacy, HLM is a small nonprofit; it has fewer than 20 full-time employees. HLM shows how a small number of employees can provide plain-language services effectively.

This chapter investigates how HLM integrates health literacy and plain language to do ethical work. It describes how HLM content and services address BUROC situations, who creates plain-language content at HLM, how HLM provides its services, and how organizational culture and ethics affect HLM's work. The chapter closes with lessons that plain-language professionals can take away from HLM.

Background of Health Literacy Missouri

HLM launched in late 2009 to address the problems of low health literacy in the state. On its website, HLM identifies several problems that affect the health of

Missourians. Compared to other states, Missouri has high rates of premature deaths, obesity, smoking, and preventable hospitalizations. Officials estimate that 1.6 million Missourians struggle with low health literacy (around one-quarter of the state's estimated population). Low health literacy correlates with many of the health issues Missourians face. People with low health literacy are more likely than others to be hospitalized, more likely to use emergency rooms, less likely to follow through with their treatment plans, and less likely to obtain preventative care (Health Literacy Missouri 2014c).

HLM provides many services to clients; these services fall into three main categories. The first category is training. HLM offers a number of training courses. Some are hour-long workshops on health literacy while others are longer sessions that include role-playing lessons on physician–patient communication or instruction on creating clear, readable materials for patients. HLM's annual summit provides panel discussions and workshops on health literacy and a venue to honor health-literacy scholars and practitioners. The second category is health-environment assessments. These assessments look at patient encounters from the patient's point of view. They consider everything from intake processes and a facility's signage to discharge processes and the documents patients receive. HLM helps clients apply assessment results to policies, clinical practices, and the design of health systems (Health Literacy Missouri 2014a). The third category is plain-language review of documentation. HLM's plain-language specialists analyze documents from clients to determine how successfully they apply plain language and to identify areas for improvement.

The current CEO of HLM is Dr. Catina O'Leary, who has a doctorate in social work and a background in research. A board of directors oversees HLM and advises the CEO. Its members, from cities and towns across Missouri, include medical professionals from a range of specialties, community members, and nonprofit advisors.

Citizens with low health literacy regularly face BUROC situations. The bureaucratic nature of many health encounters often forces patients into unfamiliar territory. Said Megan Rooney, manager of plain-language programs at HLM, "Often, patients feel overwhelmed or intimidated by unfamiliar words, concepts and interactions" because health and medical personnel often use jargon, policy-speak, and other complicated language (pers. comm., February 21, 2014). Jenna Eichelberger, a plain-language editor, said patients often feel confused: "There are many policies and procedures that must be followed in health-care settings in order for patients to get the care they need. Navigating this system can be overwhelming for everyone, especially when we're facing an illness or the illness of a loved one.... Facilities are also confusing and make it difficult for patients to efficiently receive care" (pers. comm., April 16, 2014).

Both Rooney and Eichelberger said that HLM's work centers on patients' rights to understand and participate in shared decision making with their health-care providers. Said Eichelberger, "We believe patients should have a choice when it comes to their health and that health care providers need to respect that choice" (pers. comm., April 16, 2014).

Many health and medical encounters are critical. Rooney said that misinterpretations and miscommunications lead to poor health outcomes and medical errors (pers. comm., February 21, 2014). Eichelberger said that many health-care conversations involve making important decisions quickly: "These conversations can be confusing and sometimes scary for patients. HLM works to make these conversations less daunting and more informative, allowing patients to be more confident in their decisions regarding their own health" (pers. comm., April 16, 2014). HLM's approach of working not only with patients but also with health-care providers throughout the state is noteworthy. HLM empowers patients with knowledge they can use in conversations with providers, and HLM counsels providers to better understand patients' points of view and their levels of health and medical knowledge.

Personnel Who Create, Edit, and Teach Plain-Language Content

While HLM does create some plain-language materials for clients, its main services include reviewing and editing documents to improve clients' use of plain language as well as training clients to communicate clearly in writing and in conversations with patients. Rooney has the longest tenure in the plain-language group and manages its activities. She earned a master's degree in education, which she called incredibly helpful to her plain-language work. She said that her studies around health behavior theory provided "the framework to understand the issues related to improving community health through improved communication" (pers. comm., February 21, 2014). Eichelberger graduated in 2013 with a bachelor's degree in health science. She has also earned the Certified Health Education Specialist credential (pers. comm., April 16, 2014). In the summer of 2014, Lisa Cary joined the team as a plain-language writer after a long career in the insurance industry and work as a freelance communicator. Cary helps with HLM's insurance-literacy initiatives and other projects (pers. comm., August 6, 2014). Others who participate in HLM's communication work and training have master's degrees in fields such as public health, journalism, and business administration.

Practices and Processes for Creating, Editing, and Teaching Plain-Language Content

HLM has several practices in place to successfully create and edit plain-language materials and to train others to use plain language and follow effective health-literacy principles. These include reviewing documents systematically, evaluating readability and plain language in more than one way, training and empowering others to use plain language, understanding the challenges affecting audiences with low health literacy, and assessing the systems in which patients receive documents. As with many plain-language practices, several of these at HLM overlap and reinforce each other.

Review Clients' Documents Systematically

Over the course of a few years, HLM created its own manual to guide its plain-language work. Much of this work involves reviewing documents from clients and advising clients on applying principles of plain language and health literacy more effectively. Rooney said the manual, about 50-pages long, is "a tool full of evidence-based principles to assess health content and documents for clear language and design. Much of the research inside the manual comes from the field of adult learning. It is designed to help writers develop content that is appropriate for people with limited literacy" (pers. comm., February 21, 2014).

In 2014, HLM developed its own proprietary review checklist out of the evidence-based guidelines in the plain-language manual. While HLM still refers to the manual for background information, writers and editors use the checklist more frequently. The checklist focuses on several aspects of health literacy and plain language. Said Eichelberger, "The checklist allows HLM to standardize our recommendations and give our clients a clear review that follows plain language principles. It focuses on helping writers create content for people with limited literacy skills. We review health materials for purpose, organization, design, navigation, plain language, readability, and interactivity" (pers. comm., April 16, 2014). Eichelberger described interactivity as encouraging active participation with content, such as answering a question or writing a list. In health literacy, interactivity helps keep a reader engaged (pers. comm., May 1, 2014).

Evaluate Readability and Plain Language in More than One Way

HLM approaches readability and plain language systematically, but at the same time it does not do so with a purely formulaic approach. A document is not necessarily plain or readable because it shows certain characteristics at the surface level. As Rooney said, "Readability tools are just one piece of the puzzle. They're a good way to assess at a glance where your materials are, but these tools shouldn't be relied upon as the sole measure of clear communication (pers. comm., February 21, 2014).

Eichelberger said that HLM uses Flesch-Kincaid readability assessments and SMOG scores in its reviews to give clients a general idea of the reading level of their materials. McLaughlin's (1969) SMOG grading method involves counting the number of polysyllabic words in a set of sentences. Yet, Eichelberger said, HLM acknowledges the limitations of these methods: "Readability assessments are not always accurate as they do not assess the organization of information or layout" (pers. comm., April 16, 2014). Using the organization's proprietary checklist, HLM also evaluates the effectiveness of navigation features, organization, and document design to develop a thorough understanding of a document's strengths and weaknesses.

Train and Empower Others to Use Plain Language

Training is an essential part of HLM's mission to improve the health of Missourians by closing the gap between patient skills and the demands of the health-care system. HLM trains doctors and other health professionals to communicate better with patients by using principles of plain language and best practices for health literacy. Said Rooney, who teaches some of the training courses, "We believe that empowering health care organizations to improve the health literacy in their own environments will ultimately reach more people than HLM could alone" (pers. comm., February 21, 2014). Eichelberger added that training others "is an integral part of health literacy" (pers. comm., April 16, 2014).

Understand the Literacy Challenges Affecting the Audience

While principles of plain language include carefully analyzing the audience and testing documents with audience members, the complexities of low health literacy require professionals to deeply understand readers' challenges, fears, and concerns. Eichelberger said, "It is important to understand plain language, but without understanding other aspects of health literacy, it will be difficult to make broad improvements in the community" (pers. comm., April 16, 2014). She said that professionals must understand health-care systems, medical providers' environments, and cultural competency to reach audiences with low health literacy effectively.

Said Rooney, "Understanding plain language is a start, but it is only one technique that can help address health literacy. I find that keeping up on the science of the field as a whole makes me a more thoughtful, informed plain language editor" (pers. comm., February 21, 2014). The field of health literacy develops continually. Changes in medical knowledge and best practices affect health literacy as do regulations and laws about health insurance, and developments within specific communities—from diseases and natural disasters to educational programs and galvanizing community events—affect health literacy on local levels. HLM staff need to learn about health literacy continually in order to serve their audiences effectively.

Assess Not Just Documents but Also the Systems in which Patients Receive Documents

The brochures, handouts, and discharge instructions that patients receive are parts of the larger systems in which patients receive treatment. The procedures used in medical offices and even the physical layouts of spaces can affect patients' health literacy—and contribute to the stresses they feel in BUROC situations.

The HLM health-environment assessment team enters and explores clients' facilities just as patients do. Eichelberger said this process can identify health-literacy barriers that patients may face, such as finding their way into facilities or getting through phone trees to speak to actual people (pers. comm., April 16, 2014). Rooney said the team starts by documenting the best practices already used in the facility; it then identifies any barriers to health literacy. Frequently, the team recommends improving the written materials provided to patients (pers. comm., February 21, 2014). Eichelberger said that the work of the health-environment assessment team complements plain-language work by identifying other aspects of health literacy that improve patient understanding and provider–patient communication (pers. comm., April 16, 2014).

The improvements to health literacy provided by the health-environment assessments complement HLM's training for clients on writing and speaking in plain language. Sometimes the environment and the policies contribute to health-literacy problems. Without an awareness of the environments within which providers work, HLM cannot fully address health literacy for patients of a particular clinic or health-care system. By integrating plain-language services with environment assessments, HLM can bring more of its collective expertise to bear on particular problems.

Organizational Culture and Plain Language at Health Literacy Missouri

Health literacy is the chief concern for HLM, but plain language is a primary means of improving and promoting health literacy. Because those concerns relate so closely, plain language is part of the culture at HLM. As plain language receives more attention from politicians and policy makers in the United States, more organizations will recognize the benefits of plain language. Rooney said, "There is a great demand for plain language services in the community, especially since President Obama signed the Plain Writing Act in 2010. Plain-language editing and translation is known in the field as one way to directly, tangibly, and positively impact the health community" (pers. comm., February 21, 2014).

Linking Ethics and Plain Language at Health Literacy Missouri

Plain language complements not only health literacy but also HLM's desires for ethical action. Eichelberger said, "Health Literacy Missouri is very concerned about health-care ethics. Our goal is to help people make good health decisions every day. We believe it is unethical to expect a person to make good health decisions without fully understanding the information given to them" (pers. comm., April 16, 2014). Rooney said that HLM's core mission centers around health-care ethics. Like many other health-literacy advocates, HLM believes a patient's literacy level should not be an excuse to justify a lack of understanding about medical conditions or treatment options (pers. comm., February 21, 2014).

An online video describing HLM's plain-language program also affirms the link that HLM sees between plain language and ethics. This video includes an interview with Dr. Rima E. Rudd, a senior lecturer at the Harvard School of Public Health. Speaking of professionals who write health and medical materials, Rudd said, "We should be writing our materials to match the skills of the people who are members of our intended audience." Rudd points to several thousand peer-reviewed articles that tell us that a large amount of health information is inaccessible because members of the target audience cannot understand it: "If we continue producing that kind of material that is inaccessible," said Rudd, "my claim is that it's unethical" (Health Literacy Missouri 2014b). The review of technical communication literature on ethics in chapter 2 showed that professionals consider it unethical to mislead an audience deliberately. Rudd, however, finds it unethical for medical professionals to fail in attempted communication when the content is not audience appropriate. The ethical burden that Rudd and other plain-language advocates see is part of a dialogic approach to ethics.

Supporting Dialogue with the Audience

I asked Rooney and Eichelberger which of Buber's communication models, I–It or I–You, they thought best described the view of HLM's relationships with its audiences. Both said HLM prefers the dialogic I–You model. Rooney said, "HLM believes that health literacy is not the sole responsibility of the patient, but the provider as well. We believe that good health care is always a dialogue, not a presentation of facts" (pers. comm., February 21, 2014). Said Eichelberger, "We focus our work on incorporating the audience, instead of simply talking to or at the audience. HLM believes that respect for the audience is integral to patient empowerment" (pers. comm., April 16, 2014). Respect is an integral component of Buber's "narrow ridge" discussed in chapter 2. The narrow ridge is a place where two parties separated by entrenched, deeply seated differences can meet and share common ground. HLM empowers patients to learn and understand more about the health challenges they face. HLM also trains health-care providers to better understand situations from patients' points of view. Through its work, HLM supports ethical dialogue about health literacy and health-literacy discussions between patients and providers.

Conclusion: Key Takeaways about Plain Language from Health Literacy Missouri

HLM promotes health literacy in Missouri and trains others to adopt policies and procedures that will improve health literacy throughout the state. HLM not only creates materials in plain language but also trains clients to use plain language. The ethical principles HLM applies include respecting individuals' rights to make health-care decisions, a feminist awareness of the imbalance of power between patients and health-care bureaucracies, and the feminist ideal of care, which it enacts by training others to better serve those with low health literacy. From HLM, three lessons emerge that can benefit other organizations:

1. **Offer services that link the organization's strategic goals with clients' needs.** HLM could have chosen to promote improved health literacy by focusing on creating health and medical documents for patients; it could have become a vendor to whom organizations outsource their content development. Instead, HLM primarily provides training and expertise to empower medical providers and other organizations in Missouri to create more effective materials themselves. HLM's services create greater knowledge and application of plain language and health literacy throughout the state.

2. **Go beyond the words on the page: understand how the audience develops and experiences health literacy.** HLM goes beyond the words on the page in two main ways. First, HLM monitors developments in health literacy, especially for low-literate populations. Second, HLM conducts health-environment assessments. The systems and facilities through which patients receive care can directly affect how and whether they understand the information they receive. Both of these activities complement HLM's expertise in plain language and help provide deep knowledge about audiences for health and medical information.

3. **Record institutional knowledge about plain language and pass it on.** HLM developed its own manual on plain language to guide its reviews of client documents. HLM later adapted this manual into a proprietary checklist that its editors and writers can use. By taking steps to codify this knowledge and maintain its in-house use, HLM's writers and editors have been able to work more systematically and consistently.

HLM focuses on the rights of citizens and patients to make decisions about their health and their health care. As a resource for the state of Missouri, HLM knows that poor health literacy contributes to several health problems that are prevalent throughout the state. These problems, such as high rates of obesity and high rates of hospitalizations, are critical in financial terms as well as in consequences for the lives of individual Missourians. Low health literacy means that important concepts of health and medicine are unfamiliar to millions of people—an estimated 1.6 million of whom live in Missouri. Rather than treating these individuals as mere data points in a set of poor results, HLM strives to help them, to help them help themselves, and to help others help them. HLM shows a Kantian respect to individuals who often receive little of it. Rather than developing programs to coerce Missourians into better behaviors (e.g., Dragga 2011), HLM fosters dialogue (Buber 1970) among patients, care providers, insurance companies, and other interested parties.

HLM works within organizations, from large health-care systems to local health clinics, to improve individuals' health literacy and to raise awareness of health-literacy problems. HLM helps health-care bureaucracies to have better dialogue with the patients they serve. Through training programs, assessments, and plain-language review services, HLM directly addresses the challenges of low health literacy in Missouri. HLM creates narrow ridges (Buber 1965) or meeting points between Missourians with low health literacy and those who work with them. This work increases Missourians' familiarity with health and medical issues and makes patients' interactions with health-care systems less bureaucratic and more personal.

Questions and Exercises

1. What does your organization do to "go beyond the page" to understand its audiences? Do you conduct usability tests? Do you conduct interviews or site visits? What insights from HLM's activities could you apply? Write a memo of 200 to 300 words that identifies your strengths in audience analysis and identifies opportunities you have to improve.

2. What does your organization do to empower other people to understand and apply principles of plain language? Do you offer training to clients? Do you offer lunchtime seminars to other people within your organization? Think about ways your organization can promote plain language, either within the organization, within your professional community, or within your local community. Write a memo of 200 to 300 words that identifies ways your organization can inform, train, and even empower others to use plain language.

9

PROFILE—KLEIMANN COMMUNICATION GROUP AND TILA–RESPA DOCUMENTATION

In the United States, two main laws affect the documents that consumers receive when buying a house or refinancing a mortgage. Congress passed the Truth in Lending Act (TILA) in 1968 to show consumers the costs of credit in a mortgage. The Board of Governors of the Federal Reserve System, also known as the Federal Reserve Board or FRB, was responsible for TILA. The Real Estate Settlement Procedures Act (RESPA) followed in 1974 to direct the Department of Housing and Urban Development (HUD) to give consumers timely and helpful information about real-estate settlement costs (Kleimann Communication Group 2012, 1). The Truth in Lending form, or TIL, provides the costs associated with each mortgage. The Good Faith Estimate (GFE) identifies costs that the buyer must pay at settlement, or closing, while the HUD-1 form identifies final settlement costs. Consumers get one set of documents on application (an initial TIL and GFE) and a different set at closing (the final TIL and HUD-1).

Responding to the US financial crisis between 2007 and 2010, Congress passed the Dodd–Frank Wall Street Reform and Consumer Protection Act, or Dodd–Frank Act, in 2010. Dodd–Frank transferred responsibility for TILA and RESPA to a newly created organization, the Consumer Financial Protection Bureau (CFPB). The Dodd–Frank Act required the CFPB to propose updated rules and to integrate the disclosure forms required under TILA and RESPA for mortgage-loan transactions covered by those laws. The integrated disclosure forms help lenders comply with TILA and RESPA; by using plain language instead of technical language, they help consumers better understand mortgage transactions (Kleimann Communication Group 2012, 2). Kleimann Communication Group won a contract to work with the CFPB on two types of mortgage disclosures: loan estimates and closing-cost disclosures. Kleimann Communication Group started work on the TILA–RESPA project in early 2011. Generally, the public does not see the evolution of government documents during a project like this. But the CFPB informed the public throughout the process and accepted public comments. The CFPB released the prototype disclosures in July 2012 in a proposal for a new rule about mortgage disclosures. Kleimann Communication Group conducted additional tests on the disclosures in 2012 and 2013. The CFPB released its final TILA–RESPA rule in November 2013. The new rule and the new disclosures take effect in August 2015 (Consumer Financial Protection Bureau 2013a).

Kleimann Communication Group provided the following list of three objectives for this project that it and the CFPB shared:

- *Comprehension.* The disclosures should enable consumers to understand the basic terms of a loan and its costs, both immediate and over time.
- *Comparison.* The disclosures should enable consumers to compare one loan estimate to another and identify the differences. The disclosures should also enable consumers to compare the loan estimate to the closing disclosure so that they can identify differences between the two and understand or ask about the reasons for those differences.
- *Choice.* Both comprehension and comparison should enable consumers to make informed decisions. For the loan estimate, consumers should be able to decide on the best loan for their personal situation. For the closing disclosure, they should be able to decide whether to close on the loan after reviewing the final terms and costs. (Kleimann Communication Group 2012, 2–3)

This chapter profiles the TILA–RESPA project because thousands of people, if not millions, will use the revised documents. In addition, a small contracting firm completed the project; many plain-language professionals work as independent contractors or in small firms similar to Kleimann Communication Group. While most projects will not have as great a reach as the TILA–RESPA revisions, the processes that Kleimann Communication Group used are useful and informative for other plain-language professionals. Another noteworthy aspect of this project is the recognition it received from the Center for Plain Language. The loan estimate form received the 2014 ClearMark Grand Award in addition to a ClearMark for the best revised document from a government agency (Center For Plain Language 2014).

This profile shows how Kleimann Communication Group used plain language, iterative processes, and usability testing to create clear, usable forms for consumers and professionals involved in purchasing homes. The chapter describes how the TILA–RESPA project addresses a BUROC situation, how Kleimann Communication Group completed the project, and how organizational culture and ethics affected the TILA–RESPA project. It closes with lessons that plain-language professionals may take away from the TILA–RESPA project.

Background of Kleimann Communication Group

Susan Kleimann has led Kleimann Communication Group since 1997. Based in Rockville, Maryland, Kleimann and her staff have worked on many projects for the federal government as well as firms in private industry. Kleimann has a strong background in plain-language communication and the usability testing of documents. After some years of teaching English and history to students in junior high school, Kleimann pursued a PhD in English with an emphasis in rhetoric and composition and taught college writing courses. Kleimann won a fellowship to work at the General Accounting Office (the GAO, now called the Government Accountability Office). Her doctoral dissertation analyzed how group culture affected the GAO's process for revising reports. Kleimann said her work, from teaching junior high through working at the GAO, had focused on helping writers to think clearly and logically as they wrote (pers. comm., November 25, 2013).

After completing her doctorate, Kleimann became director of the Document Design Center (DDC) at the American Institutes for Research. (Chapter 1 describes some of the DDC's work. Dana E. Chisnell, who was featured in chapter 5, worked there before Kleimann's tenure.) Kleimann said her experience at the DDC was transformative in shifting her focus from writers to users. She learned new tools and techniques for actively engaging with readers and observing them. Through processes of testing, she said, "we were in a position of seeing when readers used documents correctly and when they didn't use them correctly. It was even more important when we began to understand why they were using the document correctly or what we had done wrong that kept them from using it correctly" (pers. comm., November 25, 2013). The testing processes Kleimann used at the DDC have had practical value in her consulting work. Kleimann left the DDC to start Kleimann Communication Group. She is a founding member of the Center for Plain Language and a member of the International Plain Language Working Group, a group developing international standards for plain language.

TILA–RESPA Content and BUROC Situations

The purchase of a home is a complicated process that can have long-term consequences. Some of these consequences are financial: down payments, closing costs, and mortgage payments are significant. Other consequences affect the home buyer's quality of life: the location of the home, the local schools, the amenities nearby, and even the opportunity to paint, decorate, and remodel the property to suit personal tastes. The purchase of a home involves some amount of uncertainty and risk as well. Home values can fluctuate and neighborhoods can improve or decline.

The purchase of a home—a process governed by TILA, RESPA, and other laws—certainly counts as a BUROC situation. The purchase of a home is an undeniably bureaucratic event. Said Kleimann, "I think it may be difficult to find a more bureaucratic situation, or a bureaucratic situation with more levels, than going into a home purchase" (pers. comm., November 25, 2013). Kleimann noted that so many groups are involved with the purchase of a home, in big and small ways: banks, mortgage agents, real-estate agents, title companies, settlement agents, and more. Each group adds its own policies and procedures, works with its own sense of ethics, and tries to add value while earning revenue. Kleimann pointed out that consumers will probably never read TILA and RESPA, but they will encounter those laws through a bureaucratic process that manifests through the disclosure forms (pers. comm., November 25, 2013).

The process of buying a home is unfamiliar for many consumers. Kleimann pointed out that mortgage lenders and title companies have their own jargon for doing business, and they are familiar with the processes they control. She added that workers at these organizations might be impatient with consumers, lacking both the desire and the incentive to help them understand all the jargon and policies. While testing drafts of the disclosures, the Kleimann Communication Group team heard from consumers who had felt rushed at closing by settlement agents who wanted them to sign contract documents quickly, without reading them closely. Kleimann said the complexity of the process almost ensures that consumers will struggle to become familiar with it (pers. comm., November 25, 2013). Because home purchases tend to be both unfamiliar and infrequent, documents in plain language will be especially helpful to consumers.

Consumers have rights to purchase homes. They have rights to choose mortgages and properties that they can afford, and clear information can help them make better choices. Kleimann described a continuum of information that progresses from fact to data to information to knowledge. Kleimann said her team's goal was to make people knowledgeable enough that they can make informed choices. Even adjustable-rate mortgages, which critics maligned during the 2008 mortgage crisis, can be good choices in certain situations: "I think in choosing a home or choosing a house and choosing the mortgage, there really isn't a single right answer for any individual. There are multiple options there, and we want to make the information clear enough. And I do think that is my ethical goal to make the information clear enough that they can make a choice knowledgeably" (pers. comm., November 25, 2013). By linking clear communication with ethics, Kleimann echoes a view discussed in chapter 2 that many professional technical communicators endorse (e.g., Walzer 1989a; Riley 1993; Shimberg, 1989a).

Home purchases are critical events as well. Consumers sign binding contracts and agree to make regular payments for the full term of the mortgage—even if they lose their jobs and even if they later disagree with some of the terms, such as higher interest rates or balloon payments. Consumers will likely experience stress if the housing market declines and affects their home equity. On the other hand, a home that increases in value can benefit a family's financial well-being substantially.

Personnel Who Create Plain-Language Content for Kleimann Communication Group

As president of her firm for almost two decades, Kleimann has developed a small, tightly knit team. Other individuals may join projects when the project scope warrants it. Kleimann said that she and three other team members represented Kleimann Communication Group on the TILA–RESPA project. While writing ability is important for plain-language work, Kleimann identified two other abilities that are especially important to her. The first is the ability to hold and test a hypothesis when creating effective documents:

> What you're trying to do is go out there and look at that hypothesis to see if it works. And if it works, you have to question yourself about whether or not it really works or you just wanted it to work. And if it didn't work, you have to question yourself about why it didn't work. So I want to see that willingness to stay in that suspended state of "It's an idea, we're just trying it. It might work, it might not work." What's important is that we get to a place that does work. I think this [being able to investigate a hypothesis] is maybe the most critical characteristic to have. (pers. comm., November 25, 2013)

The second ability that Kleimann desires is to be able to collaborate effectively. To Kleimann, effective collaboration involves balancing confidence with humility. Kleimann does not want her team to be the experts who charge in and tell clients what to do and how to do it. Instead, she wants them to be willing to ask questions, be tenacious while testing hypotheses, and accept insights from anyone involved on a project (pers. comm., November 25, 2013). Kleimann Communication Group collaborated with several representatives from the CFPB on the TILA–RESPA project. Kleimann and her team did not solve the problems with TILA and RESPA disclosures themselves, but they developed solutions collaboratively.

These two abilities that Kleimann values in her employees strongly support effective dialogue. To hold and test a hypothesis effectively, a person must be willing to ask questions of collaborators openly. Thus, conversation can be an important research tool. Dialogue is also a key component of collaboration. Telling clients "what to do and how to do it" is not a dialogue; that approach treats clients as Its and not Yous. Talking with clients and exchanging ideas with them is the kind of true dialogue that Buber (1970) describes.

Practices and Processes for Creating Plain-Language Content on the TILA–RESPA Project

Kleimann Communication Group thoroughly documented its processes and practices used on the TILA–RESPA project (Kleimann Communication Group

2012). The major phases of content development included context setting, formative development, and iterative development through usability testing. The public-comment period following the release of the proposed disclosures also yielded information that informed the final designs. Additional testing and final revisions followed public comment.

Context Setting

In the first phase, Kleimann Communication Group met with CFPB personnel to understand the full background of the project, to understand existing research relevant to the project, and to establish the technical content that the disclosures must contain. The team did this through three concurrent activities: reviewing relevant research, discussing technical content with experts, and brainstorming through the Kleimann Blank Sheeting Process® (Kleimann Communication Group 2012, 27).

The Kleimann Blank Sheeting Process involved a series of steps to identify and set priorities for the tasks that consumers must accomplish with the disclosures and to identify the information needed for each task. The process helped team members clarify a disclosure's purpose, identify key design challenges, identify key tasks performed with the disclosure, and discuss characteristics of key populations who would use the disclosure (Kleimann Communication Group 2012, 27).

Although the CFPB had provided a review of technical research on the disclosures, Kleimann Communication Group reviewed other relevant research on both content for the disclosures and design features that improve use of the disclosures. This research process included reviewing the list of statutory requirements for the disclosure and conducting a meta-analysis of the existing academic and practitioner research in order to establish the essential information needed to enable consumers to make decisions and compare loan offers. Kleimann Communication Group also analyzed background materials from the CFPB and researched sources to document current issues with the TIL, GFE, and HUD-1 disclosures (Kleimann Communication Group 2012, 27–28).

In the technical content meetings, the team combined the information from brainstorming and from reviewing relevant literature. These meetings helped identify the required content of the disclosures, reasons behind the placement of text and design elements, and emerging issues to consider about the disclosures (Kleimann Communication Group 2012, 28). The time and energy Kleimann Communication Group spent understanding the context of mortgage documentation provided an important foundation of knowledge to guide the latter phases of the project. This deep foundation allowed Kleimann Communication Group to better understand the needs of mortgage professionals and related real-estate personnel while it also supported the group's understanding of how consumers will read and use the mortgage documents.

Formative Development

In the second phase, project team members began to apply the knowledge that they gained from context setting. The team first created user personas to help envision the mortgage disclosures from consumers' points of view. Next, the team used rapid prototyping to develop models for the disclosures and to incorporate informal feedback quickly. Personas and rapid prototyping are common elements of user-centered design processes (Kleimann Communication Group 2012, 28–29).

The personas represented a range of fictional individuals with traits that are common among actual members of the target audience. These traits included life circumstances, such as how long a person might expect to stay in a house; age and ability, including abilities to compare options and to read small print on a page; quantitative literacy, such as the ability to understand interest rates; financial literacy, including the ability to distinguish one-time payments from monthly payments; and visual literacy, such as the ability to interpret a graph with multiple variables. The project team made specific design choices to complement or compensate for characteristics in the personas and the target audience (Kleimann Communication Group 2012, 42–43).

Through rapid prototyping, the team used a cycle of designing and informal testing to sort through more than 100 potential designs. A group of five volunteers provided feedback on the prototypes. While this process was not scientific (and was not intended to be), it was systematic, and it brought in valuable user feedback from outside the project team. Prototyping in March 2011 focused primarily on the visual design of the disclosures while testing in April 2011 focused more on the content of the disclosures. The formative development phase gave the team two prototype designs to develop and refine through usability testing (Kleimann Communication Group 2012, 42–46). By choosing user-centered methods for the development of the revised mortgage documents, Kleimann Communication Group showed how organizations can enact Salvo's (2001) call for dialogic processes that help users and designers to better understand and relate to each other.

Iterative Development through Usability Testing

The third phase of development was the most extensive. Kleimann Communication Group used 10 rounds of usability testing to provide insight on how well consumers would be able to use the disclosures. Usability testing collects robust qualitative data about participants' abilities to use documents or tools instead of merely gauging their preferences through surveys or focus groups. It focuses on what participants do with a document or tool rather than what they say or perceive that they would do. Consumers' actual behavior often differs from the behavior they report when asked (Kleimann Communication Group 2012, 30), as many usability experts have learned. Beta testing, in which some users get access to a technology product before its full public release, provides less value as a form of usability testing than some might expect. Beta testing is too small, unsystematic, and late in the development process to provide valuable information about a product's usability (Dumas and Redish 1999, 24).

Kleimann Communication Group conducted usability tests in each of the nine geographic divisions identified by the US Census Bureau. The first and last rounds of testing occurred in Baltimore. Other tests occurred in a range of small, medium, and large cities. These cities included Des Moines, Iowa; Birmingham, Alabama; and Los Angeles, California. This geographic diversity helped the project team to get a sample of participants that better reflects the US population than would participants from one or two cities. The project team worked with 92 consumers and 22 industry professionals across the nine testing locations. The participants represented a range of ethnicities, age groups, education levels, income levels, and levels of experience buying homes. In addition, Kleimann Communication Group specifically recruited Spanish-speaking consumers for some rounds of testing (Kleimann Communication Group 2012, 33–36).

On one hand, 10 rounds of testing and 114 participants are large numbers—especially to people who have never done usability testing or have tested only a few participants. On the other hand, as Kleimann explained, they do not seem like large numbers when people understand the broad impact of mortgage disclosures. The disclosures will affect the ways that industry professionals handle real-estate transactions in the United States, and they will affect every consumer who takes on or refinances a mortgage (pers. comm., November 25, 2013). In practice, however, researchers rarely need to work with large numbers of users to test usability effectively. Nielsen (2000; 2012) writes that as few as five users are enough to identify most usability problems. He notes that when there are two or more distinct groups of users, usability experts should test a few members of each user group (2000). The project team tested between seven and twelve consumers and two to six industry professionals at each site (Kleimann Communication Group 2012, 32). Because the team worked with a diverse group of participants and used iterative design to make incremental improvements to the documents, Kleimann believes her team's testing was sufficient for identifying the disclosures' most important issues (pers. comm., November 25, 2013).

Kleimann Communication Group developed specific research questions for each group with whom they worked and for each document they tested. Research questions for consumers focused on their ability to understand loan costs, compare options, and choose a mortgage effectively. Questions for industry professionals focused on how well they understood the documents, when and where they expected consumers to struggle with the documents, and how CFPB might help professionals make smooth transitions to updated disclosures. Kleimann Communication Group interviewed each participant individually. All participants received an assigned task; team members asked them to speak their thoughts aloud while completing the task, a process called a think-aloud protocol. Interviewers also used a structured set of questions to supplement the think-aloud protocol. Finally, interviewers asked participants to compare three or more mortgages; the team wanted to understand the rationale behind trade-offs users made. The team used a method called grounded theory to analyze the results of each round of usability testing; this inductive method identifies patterns in a set of qualitative data (Kleimann Communication Group 2012, 30–33), such as users' dislike for a particular graphic on a website or their enjoyment of an intuitive feature in some software.

The first five rounds of testing focused on developing the design and content of the initial mortgage disclosure while the next two rounds focused on the design and content of the closing-cost disclosure. The final three rounds focused on refining the designs of both disclosures and on the interaction between the two disclosures (Kleimann Communication Group 2012, 37). Kleimann Communication Group used a similar structure for each round of testing. In each round, the team established a rotation for order in which participants saw the various designs to prevent the order from affecting the results. When test participants see the same documents in the same order, the results might show a primacy effect, in which participants prefer the first document they see, or a recency effect, in which participants see several items and prefer the last one they see (Colton and Covert 2007). The team used a consistent slate of interview questions. The report for each round of testing summarizes findings by categories, identifies findings that relate to design, and identifies revisions made to each prototype document in preparation for the next round of testing (Kleimann Communication Group 2012, 47–272).

Public Comment

During the usability-testing phase, CFPB published draft versions of the disclosures on its website while Kleimann Communication Group conducted usability testing. CFPB published the drafts that Kleimann Communication Group tested in each city. Consumer advocates, consumers, and people from industry submitted around 27,000 comments (Consumer Financial Protection Bureau 2013b) through a process managed by the Federal Register. Kleimann said this public-comment phase allowed people to follow the development of the disclosures and to know what to expect when the disclosures would become official. The project team considered these comments and acted upon some of them as they continued usability testing (pers. comm., November 25, 2013). Rather than using authority to take a coercive stance toward those who must follow TILA and RESPA in the course of a home purchase, CPFB and Kleimann Communication Group used a more humane, dialogic approach (Dragga 2011) by inviting public comment.

The US government invites public comment on all proposed changes to federal regulations. The proposed new TILA–RESPA rule, including the disclosures, was available for public comment between July and November of 2012. While

Kleimann Communication Group did not manage this process, the project team did receive the comments that stemmed from the process. Kleimann said that three main ideas about the disclosures emerged from the public comments. First, commenters wanted to see disclosures in Spanish. The team had already started to work on them, but testing showed that the Spanish disclosures needed more work. For instance, Kleimann said the team struggled to convey "balloon payments" in Spanish because that is not a familiar concept in Hispanic culture. Second, commenters from the real-estate industry said that the disclosure for refinanced mortgages did not work effectively; the form needed to show more clearly whether consumers would receive money back or owe money upon refinancing. Third, consumers thought the closing costs on the first page of the updated closing disclosure needed to attract more of the readers' attention (pers. comm., November 25, 2013).

Revisions to Proposed Disclosures

After the CFPB issued the proposed TILA–RESPA rule and disclosures, the project team revised the disclosures with information from the public comments and also created disclosures in Spanish. Kleimann Communication Group conducted seven more rounds of usability testing to identify final changes for the disclosures. The group also bid for and won a government contract to conduct quantitative validation testing to compare the effectiveness of the revised disclosures with that of the original disclosures. The validation testing showed that consumers could use the new disclosures significantly more effectively than the old disclosures (Kleimann Communication Group 2013b).

Organizational Culture and Plain Language on the TILA–RESPA Project

Kleimann Communication Group has worked on many projects involving government documents that the public uses. The work to create these other documents was similar in type to the work on TILA–RESPA, if not similar in scope. All of these projects involved writing in plain language and testing document drafts with readers from the target audience to make the documents as effective as possible. The principles of plain-language practice are part of the culture at Kleimann Communication Group.

Kleimann said that the CFPB, the sponsor of the project, shared her team's goal of creating effective documents as well as a commitment to do the job well. Kleimann made a point to say how much she appreciated the CFPB's support for testing documents with members of the target audience:

> I do think it was extraordinary that we had an agency that was willing to do a project of this scope with this much consumer testing. I personally and

professionally do not believe that all projects deserve this kind of testing or this intensity of testing. . . . We do lots of projects where we're not doing this scope of testing. But I think the CFPB really should be commended for being willing to do something on a project that has this kind of impact and scope. They deserve the kudos for that. (pers. comm., November 25, 2013)

This statement indicates how much Kleimann Communication Group values user testing as a part of developing plain-language documents. It also shows how the team worked with CFPB to develop and approve a plan that provided adequate user feedback; the extensive testing required CFPB to invest more time and resources than a plan with less testing would have.

By interacting with so many audience members for the mortgage-cost disclosure and the closing-cost estimate, Kleimann Communication Group used a dialogic process for content development. Kleimann described her team's work as using Buber's I–You mode of respectful communication with the audience rather than the I–It mode of communication at the audience:

What we were about was the consumer. This wasn't treating the consumer as a thing. This was treating consumers as people who had real decisions to make who lived in a real world, who knew their personal situations far better than we did. We attempted to make the forms clear enough that consumers would be able to use them to be able to make a decision that was appropriate for them. (pers. comm., November 25, 2013)

Rather than simply telling users to accept and use the forms, Kleimann Communication Group, through extensive testing with not only consumers but also professionals who handle home purchases, asked users if documents were effective and made changes in response.

Part of the team's sense of ethics involved being aware of personal preferences and trying to focus on facts over feelings. One example involved testing with consumers' information about adjustable-rate mortgages and negative amortization loans—loans that carry risks that traditional mortgages do not have. Many consumers who tested draft disclosures expressed strong reservations about such loans, but Kleimann added that these loans can provide advantages to certain home buyers—usually those who do not expect to keep a home several years. Kleimann said her team tested these potentially risky mortgage options to ensure that consumers had enough clear information to reason through decisions about them. A second example involved the amount of financial detail to include on the mortgage disclosure. Kleimann said her initial response was to combine certain fees together rather than identifying all of them. She thought consumers would prefer seeing fewer, larger numbers rather than a full list of all the costs. Given the choice between grouped costs and full lists of costs, however, consumers preferred the full lists. Thus, the final mortgage disclosure reflects consumers' preferences rather than those of Kleimann or her team (pers. comm., November 25, 2013).

Conclusion: Key Takeaways about Plain Language from the TILA–RESPA Project

An extensive, well-documented plain-language project like the TILA–RESPA disclosures project offers many lessons and insights. The ethical principles applied in this project include a deep respect for individual rights, an open dialogue among the many audiences affected by home-purchase documentation, and an awareness of the differences in power between all the parties involved in a home purchase. Five key lessons about plain language emerge from this profile, lessons that other organizations can put to use:

1. **Integrate testing and iterative content development.** Plain-language experts have long advocated gathering feedback from readers and incorporating that feedback to improve a document's next iteration. Find ways to gather feedback and then use it to improve your documents, such as through rapid prototyping with a small group of volunteers or repeated usability testing with audience members.
2. **Develop and test translated documents thoroughly.** Early testing with Spanish-speaking readers showed that it would be a challenge to present certain concepts accurately in the Spanish disclosures. After the visual designs of the disclosures were stable, the CFPB devoted more resources to developing and testing Spanish disclosures. Kleimann said that certain parts of the Spanish disclosures, such as those dealing with balloon payments, presented some of the most difficult challenges in the TILA–RESPA project (pers. comm., November 25, 2013). Testing showed that the documents Kleimann Communication Group produced were successful (Kleimann Communication Group 2013a). Without testing, the Spanish documents likely would have been difficult for the Spanish-language audience to use.
3. **Accept public comment when appropriate.** Members of the public made thousands of comments on the disclosures through a process managed by the Federal Register. Many of those comments led to improvements in the documents, such as a clearer indication on the closing disclosure of whether consumers will receive money back or owe money at closing. Members of industry expressed their ideas and concerns about documents that will affect them, and consumers were able to share concerns from their own perspectives. Kleimann said that some comments expressed encouragement, some noted that the forms in development differed noticeably from the old forms, some identified parts of the forms that needed improvement, and some reiterated information that was simultaneously revealed through the iterative testing (pers. comm., November 25, 2013). Kleimann Communication

Group took all of this feedback into account as it continued to make iterative changes in response to the feedback received through usability testing.

4. **Seek feedback from all groups that use the documents.** Soliciting feedback, whether publicly or privately, from all groups affected by a particular document is critical. Kleimann Communication Group involved industry representatives in each round of usability testing. Doing so allowed people in industry to be aware of changes in documents that will affect them and also to participate in shaping those documents.

5. **Test visual design as well as verbal content.** Visual design substantially affects how users understand a document, even if the document is already in plain language (Cutts 2009). Kleimann Communication Group used both the rapid-prototyping phase and the usability-testing phase to gather specific feedback on the design of the disclosures. This information helped team members to create a visual design that effectively complements the verbal text.

As Kleimann said, "I think it may be difficult to find a more bureaucratic situation, or a bureaucratic situation with more levels, than going into a home purchase" (pers. comm., November 25, 2013). The BUROC model aptly coincides with the situation of a home purchase. Many home buyers, especially first-time home buyers, are unfamiliar with the terminology and the processes involved in these complex transactions. Consumers have rights to choose where they want to live and to enter into mortgage agreements to purchase their homes. Consumers should not be treated as mere means for mortgage lenders, real-estate agents, and others to make money; instead, they should benefit most when they purchase or refinance a home (see Kant [1785] 1969). Home purchases have critical impacts on a family's access to education, cultural and civic amenities, and a higher quality of life. Mortgage lenders should disclose all the information that consumers need to make informed decisions. Consumers should know the consequences of both meeting and not meeting the terms of loan agreements.

The forms created in the TILA–RESPA project make a complex, bureaucratic process more familiar to consumers. The language is clearer, the organization is easier to follow, and the information in the forms is generally more accessible than it had been. Through iterative development and substantial usability testing, Kleimann Communication Group used a dialogic approach (Buber 1970) to create effective documents that address the needs of consumers, mortgage lenders, and other real-estate professionals.

Questions and Exercises

1. How much do you know about usability testing? How confident are you that you could design and oversee an effective usability test? If you know usability testing well, write up a short memo of 200 to 300 words summarizing the

important steps in conducting a usability test, and include a short list of books or online resources that have helped you. If you do not know usability testing well, conduct some research to identify the important steps. Write a short memo of 200 to 300 words, and then include an annotated list of books and online sources that could help you learn more about usability testing. Each annotation should have 50 to 100 words.

2. Find a copy of the documentation for an important purchase, and focus on one page. How easy is it to understand the form? Does the form provide all the information that a consumer needs for that purchase? Write a memo that describes the strengths and weaknesses of that page of the form, and then develop the outline of a plan to improve it. The memo and outline should total 350 to 500 words.

10

PUBLIC EXAMPLES OF DIALOGIC COMMUNICATION IN CLEAR LANGUAGE IN THE TWENTY-FIRST CENTURY

The plain-language movement continues to grow. Aided by the work of plain-language advocates and of organizations such as Clarity International, the Center for Plain Language, and Plain Language Association International, the public receives more documents than ever before in clear, understandable language from government agencies and even from private businesses.

Other individuals and companies are using innovative approaches to communicate clearly and to help authors reach their audiences more effectively. These efforts have developed outside of the plain-language movement, but their creators share similar goals of effectively reaching audiences using clear verbal and visual language to support various informational and ethical goals. In many instances, these communicators use the internet both as a medium for sharing their messages and a conduit to support dialogue with their audiences.

Innovative Approaches to Clear Communication and Dialogue

This chapter describes several innovative approaches to clear communication and dialogue with audiences. Innovators include the firm Common Craft, with its trendsetting approach to online explanation videos; Booster Shot Media, which provides print materials and online videos about asthma for young audiences; John Wiley & Sons, whose For Dummies series includes hundreds of books and a robust supporting website; Mignon Fogarty, who turned her Quick and Dirty Tips podcasts into a multimedia publishing venture with Macmillan; and the Alan Alda Center for Communicating Science at Stony Brook University, which uses techniques from storytelling and improvisational theater to help scientists learn how to communicate with nonexperts.

Common Craft

Common Craft is a small but influential producer of online explanation videos operated by the husband and wife team of Lee and Sachi LeFever. Their motto is "Our Product is Explanation." Indeed, Lee LeFever wrote a book titled *The Art of Explanation* (2012). LeFever said the Common Craft approach

> fills the gap that is often left open by experts who are not interested in explaining the basics, or feel that they've already been covered. As our world grows more and more complex, the need and demand for useful explanations is growing, and we want to be there for people who just want to grasp the big ideas quickly. If we do it effectively, we can hopefully motivate them to keep learning. (pers. comm., November 11, 2013)

LeFever's comments identify the value that effective explainer videos can provide. These videos reach audiences who want to learn about new fields but lack expertise in them. Rather than dumbing-down information, videos by Common Craft and others make information accessible to nonexperts.

Common Craft creates stop-motion animated videos with low-tech, hand-drawn visuals and scripts in familiar language. These videos typically incorporate background information, simple stories, and analogies that educate, inform, and occasionally entertain. Common Craft shares the videos on its website and offers a membership service for using them in schools, training sessions, and other instructional applications. LeFever said the company's goal is to make ideas easier to understand through video explanations (pers. comm., November 11, 2013). Topics range from digital literacy and cloud computing to investing and emergency preparedness. One of the company's earliest videos, RSS in Plain English, appeared online in April 2007. Within three years, it had received more than 1.1 million views, and several organizations had licensed it for private use (LeFever 2010).

LeFever said the couple had not planned a career in making online videos. Common Craft's original focus was to consult with organizations about online-community strategy (Common Craft 2014). The couple made their first videos when YouTube was just starting to gain popularity: "We wanted to test if online video could be an asset for Common Craft. Using paper cut-outs and a whiteboard was my wife Sachi's idea and we tried the format on a whim. We had no idea where it would take us" (pers. comm., November 11, 2013). LeFever said the simplicity of the Common Craft approach facilitates its effectiveness: "Today, we see that video is an amazingly efficient way to communicate complex or complicated ideas by drawing on two senses at the same time: sight and sound. By using limited hand-drawn visuals, we can reduce a lot of the visual noise that comes with video and focus the audience on the ideas" (pers. comm., November 11, 2013). For example, the hand-drawn people in Common Craft videos do not have faces, and the web browsers in the videos have sparse, simple interfaces. The cutouts are so distinctive and popular that Common Craft gives sets of figures to paying members who want to use them in their own videos.

In its first several years, Common Craft posted videos on YouTube and also provided custom videos for clients. In 2013, Common Craft shifted its focus to providing subscription access to a library of videos that primarily concern digital literacy. LeFever said that the current audience includes classroom teachers, corporate trainers, librarians, consultants, and even start-ups. They subscribe to Common Craft for resources that help people feel confident about using technology (pers. comm., November 11, 2013).

In his book and in online articles and blog posts, LeFever has said that empathy for the audience is a key component of a successful explanation video:

> The product of a successful explanation is understanding. In order to achieve that goal, we have to think about our audience and what they already understand or know. We have to make assumptions and imagine what our words sound like to them. To be understandable, we must transform the ideas in our heads into a form that is likely to resonate with theirs. That's what it means to empathize—to imagine what it's like in the shoes of another. Without empathy, it's very difficult to adjust our language and ideas for our audience. (pers. comm., November 11, 2013)

LeFever's approach has a lot to do with dialogue. Instead of creating videos that speak at the audience, as in Buber's I–It mode of communication, LeFever uses empathy to find ways to speak to the audience in the I–You mode.

Common Craft does solicit feedback from its audience. The company provides members a tool for suggesting and voting on video titles. Common Craft frequently develops videos in response to audience suggestions. Said LeFever, "The online voting tool gives us a valuable way to understand what the members need" (pers. comm., November 11, 2013).

LeFever hopes that Common Craft videos will help people to be responsible citizens. Several of its videos identify how to do the right thing through avoiding plagiarism, communicating online, and protecting reputations online. "But," said LeFever, "for us, it's not enough to simply explain an ethical issue in terms of right versus wrong. We work to create an experience that shows why it makes sense that something is a poor choice.... The understanding that we want to create is not on the right/wrong axis so much as the makes sense/doesn't make sense axis" (pers. comm., November 11, 2013). LeFever shows again that the Common Craft approach is dialogic. If Common Craft simply wanted viewers to know what is right and what is wrong, the company would regard viewers as Its and write scripts that speak at the audience. To give the audience greater understanding, LeFever uses empathy and respect to address viewers as Yous.

Booster Shot Media

Booster Shot Media is a collaboration between health communication specialist Gary Ashwal and pediatric allergist Dr. Alex Thomas. Thomas has experience in comics, art, and illustration in addition to his medical training while Ashwal has a background in health communication and video production. Ashwal said, "We started Booster Shot Media because we wanted to combine our creative work with our interest in patient education. Our goal is to produce dynamic and helpful educational materials for patients" in an easy-to-understand and memorable format (pers. comm., December 3, 2013).

While the company offers general services in video production and website content development for medical audiences, much of its recent work is in comics for children who have asthma. Through Booster Shot Comics, the duo develops characters, stories, and visuals that are memorable to patients, providers, and the public. Ashwal said, "It's not about slapping together a cartoon because 'Hey, kids like cartoons!' We use the language of comic art, the use of characters to represent more complex concepts, and the recurrent tropes in comics to create a more intuitive learning experience for the patient. We try to judge ourselves critically and strive to work at the top of our ability" (pers. comm., December 3, 2013).

Booster Shot Comics incorporate a variety of media to engage and educate young audiences (and their parents). In addition to printed comic books such as "Iggy and the Inhalers," in which Iggy the Inhaler helps to save the day, Booster Shot offers trading cards of characters in the comics. These characters include asthma villains such as mold and cigarette smoke and heroes such as Broncho, the bronchodilator medicine that opens passages in the lungs. Online videos, including animated cartoons and hand-drawn diagrams on a white board, round out the instructional materials. Figure 10.1 shows one of the trading cards.

While comics can be a particularly engaging medium to create (and fun to produce), Ashwal pointed out they must still maintain a rigorous approach to developing informational content: "While we aim to produce material that we enjoy personally, we also want to impress both our creative colleagues and our health care colleagues equally. In that sense, we keep high standards for visual aesthetics and also strive to have content that is scientifically accurate and up-to-date" (pers. comm., December 3, 2013).

Ashwal said that their primary audience includes any person seeking health-care information. These audience members need information that is easy to understand. Ashwal and Thomas tend to create their materials for patients, especially children. At the same time, they hope their tools are useful to doctors, nurses, and other health-care providers who need to educate their patients about specific diseases or treatments (Ashwal, pers. comm., December 3, 2013).

Ashwal said that they do not view their creative processes as a dialogue with the audience, but they do solicit feedback from several groups. To get feedback, they share their early-stage work and their thoughts on creating it with patients,

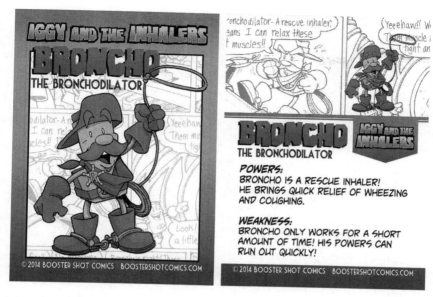

FIGURE 10.1 A Booster Shot Comics trading card for Broncho, the bronchodilator inhaler. Reprinted with permission of Gary Ashwal and Alex Thomas.

medical experts, and creative colleagues: "Right now, we just try to put our work out and see who likes it and feels compelled to respond. Then, we try to keep a dialogue going with those individuals through conferences, email, or meeting in person" (pers. comm., December 3, 2013). They frequently attend the Comics and Medicine conference in addition to other meetings focused on asthma care, and they post materials on the Booster Shot Comics blog.

Ashwal and Thomas use Facebook and Twitter to share thoughts about health-care education as well as behind-the-scenes details regarding the creation of their materials. These channels provide some degree of dialogue with users of social media. Direct feedback from patients, such as those in Thomas's clinic, also helps them test ideas and improve their materials.

Ashwal and Thomas see an ethical impact in the information their materials provide. Patients need tools and information to manage their conditions and participate in treatment decisions, and Ashwal and Thomas believe health-care providers must provide information that patients can understand. As providers have less time available to educate patients, Ashwal and Thomas hope their materials make patient education both more efficient and more meaningful: "We hope our work has an ethical impact in terms of providing patients with clear information they can understand" (Ashwal, pers. comm., December 3, 2013). Comics provide a unique and memorable way to reach young asthma patients. Ashwal and Thomas treat their readers as Yous and not Its by creating materials that are appropriate for readers in their stages of life.

John Wiley & Sons: For Dummies® Series

Since the publication of *DOS For Dummies* in 1991, the For Dummies series of books has approached challenging technical subjects with a unique, lighthearted tone and a clear, plain style. Consumers have bought more than 250 million copies of the more than 1,800 titles in the series (John Wiley & Sons 2013). In addition to its many book titles, the publisher offers Dummies.com, a portal with access to an extensive collection of how-to information on a variety of subjects.

David Palmer is brand director of the For Dummies series, which has the tagline Making Everything Easier. Palmer said For Dummies is at its best when addressing difficult or loathsome tasks—such as solving algebra problems or paying taxes—and making them easier. Palmer identified three key aspects of the For Dummies approach:

> Firstly we bring fun, humor and irreverence—even taxes can be funny, and an irreverent approach to algebra is far more likely to engage than a dour textbook. Secondly the content is modular—you don't have to have read any of the rest of the book to get what it says on page 237, so you can dip in and out to find out just what you need. Thirdly there is the variety of formats in which Dummies content is delivered—in key STEM [science, technology, engineering, and mathematics] subjects you can get a book, an e-book, a workbook, videos, practice questions, and an online practice system that will monitor and track your progress. We provide routes to success for every type of learner. (pers. comm., November 22, 2013)

Palmer added that the For Dummies approach resonates with readers who are smart enough to know they need help figuring something out. If sales figures are any indication, the company's ironic approach for attracting readers has worked well: "It's not a mark of shame to be a Dummy; it's a badge of honor. The words 'For Dummies' have entered the culture as shorthand for a guide to complex information" (pers. comm., November 22, 2013). In addition to its books, the For Dummies series provides many resources online, including articles, newsletters, summaries of key ideas from its books, and instructional videos.

Palmer said the For Dummies audience "is as broad as it is tall, but there are certain segments that we focus on" (pers. comm., November 22, 2013). One segment is personal and home computing. For Dummies addresses all major computer operating–system releases from Microsoft and Apple with a range of content from novices to advanced users. In education, For Dummies serves students in STEM subjects with supplementary materials in both hard copies and electronic media (pers. comm., November 22, 2013). Subjects covered in the For Dummies series range from art history to existentialism and from organic gardening to photography.

Palmer says that John Wiley & Sons works hard to nurture the dialogue between the publisher and the readers through social media. Wiley has one team member assigned solely to the For Dummies social media accounts and the series' general presence online. Palmer says the For Dummies brand has a strategy for ways to respond to the various types of interaction (positive or negative, feedback or discussion, conversation with the For Dummies team or about the series). Wiley also manages broader strategies for social media for purposes such as increasing sales, increasing awareness of the brand, or launching a product. Palmer said, "Listening to the dialogue is both a key route to understanding the position of the brand in our customers' views and a key to generating new content and product ideas" (pers. comm., November 22, 2013). Palmer added that Wiley strives to maintain consistency of personality and voice across all our For Dummies products and social media. Wiley wants the experience of reading *Windows 8 For Dummies* to be the same as that of engaging with For Dummies on social media (pers. comm., November 22, 2013).

Palmer sees ethical action in serving the audience effectively: "We are in the business of education, and our tagline of making everything easier is our ethical standard. We were born out of a desire to make the challenge of understanding how to use a computer more egalitarian. We challenged the dense manual that came with your computer with a healthy dose of irreverence and fun, and we continue to do so today" (pers. comm., November 22, 2013). Palmer said that For Dummies will probably help readers to make better-informed choices, but he added that making a reader more ethical probably lies outside the scope of what a For Dummies book can do (pers. comm., November 22, 2013). Nevertheless, the For Dummies series includes books on general ethics, business ethics, and health-care ethics. The For Dummies approach is unconventional but is definitely dialogic. By going out of their way to make challenging subjects understandable and even enjoyable, For Dummies authors treat readers as Yous and not Its.

Mignon Fogarty: Grammar Girl and Quick and Dirty Tips

As Common Craft showed with its online explanation videos, sometimes a simple idea grows into something much greater when internet users take notice. The story of Mignon Fogarty's creation of Grammar Girl and Quick and Dirty Tips is similar. Fogarty has a background in science writing and technical writing. In 2006, she was sitting in a coffee shop and editing some technical documents with several grammatical problems. Podcasting was a new trend, and Fogarty thought that she could create a podcast to share some of her knowledge about writing. Thus, the idea for her Grammar Girl podcast was born (Williams 2007). Grammar Girl soon became one of the most popular podcasts on the internet. After the success of Grammar Girl, Fogarty created the Quick and Dirty Tips network and launched five other podcasts.

In 2007, an editor from the publisher Macmillan contacted Fogarty to see if she might be interested in collaborating. In the years since, Fogarty has published several Grammar Girl books with Macmillan. Fogarty and Macmillan have expanded the Quick and Dirty Tips brand of content into many subject areas across several platforms. Kathy Doyle, director of Quick and Dirty Tips, said being "quick and dirty" involves presenting information in manageable chunks that readers can use to find the answer to a question or a query (pers. comm., December 6, 2013). Quick and Dirty Tips now has 15 hosts, including Fogarty, who are responsible for creating content in their assigned areas.

Fogarty said her approach to Grammar Girl and to Quick and Dirty Tips is to provide clear, simple, accurate information in a friendly and entertaining way. She believes that people learn better when learning is fun and entertaining: "When I began the Grammar Girl podcast, much of the existing grammar content was either dry and academic or snarky. I have received many email messages over the years thanking me for being friendly and helpful" in an area that often intimidates younger students and adult learners alike (pers. comm., November 24, 2013).

All the hosts in the Quick and Dirty Tips network have developed a persona that helps them to connect with the audience. Fogarty is Grammar Girl, Ben Greenfield is Get-Fit Guy, and Amanda Thomas is Domestic CEO, to name a few. And all the hosts have a web page to describe their mission for the information posted on their site. The hosts get to know their audiences well by interacting with them through email, social media, and even conference sessions; the hosts are not only creators of content, but they are people with whom audience members can interact directly. Doyle remarked that Fogarty recently attended the convention for the National Conference of Teachers of English. There, Fogarty talked to several teachers who use her materials in the classroom. Doyle said they gave Fogarty valuable ideas for new interactive content, teacher guides, lesson plans, and even software apps (pers. comm., December 6, 2013).

Hosts have considerable freedom to develop content they think will be appropriate. Said Doyle, the goal of the website is not simply to collect a lot of information about particular topics but to provide content that is "personally curated, hand crafted, and tailored to fit the mission and the audience" (pers. comm., December 6, 2013). Common forms of Quick and Dirty Tips content include audio and video podcasts from the hosts, podcast transcripts posted online, online newsletters, and even books. Fogarty recently created a grammar-quiz app for purchase as well.

Doyle said it is a challenge to define all the audiences for Quick and Dirty Tips because they are diverse. She provided two examples of this diversity. The Dog Trainer, Jolanta Benal, has a narrowly focused audience. The Get-It-Done Guy, Steve Robbins, has a broader audience for his information about personal productivity. Doyle said that part of the job for hosts is to get to know their audiences well. Because hosts know their audiences, she trusts them to create content that they think is appropriate (pers. comm., December 6, 2013). Although the audiences' interests are diverse, they share a common desire for practical, useful information.

Dialogue with the audience is important to Quick and Dirty Tips hosts. Fogarty said her communication with the audience goes both ways: "I get many of my topic ideas from my audience, and it's not unusual for my opinion or perspective to be altered by an interaction with an especially thoughtful fan. Many of my listeners and readers have a more extensive background in linguistics and education than I do, and I'm always open to their ideas" (pers. comm., November 24, 2013).

One aspect of the popularity of Grammar Girl content is that Fogarty receives numerous messages from her audience. Fogarty said she could easily spend all of her time responding to questions from Twitter, Facebook, and email. A part-time assistant helps her by responding to simple email questions before forwarding others to Fogarty. Fogarty says it is important to maintain some distance from the audience: "In the past, I found that when I answered a person's question by email, he or she then often assumed I would always be at his or her disposal or that I would be available just to chat, and I find it difficult to manage people's expectations without seeming rude. Having an assistant has been a big help in that area" (pers. comm., November 24, 2013). Figure 10.2, from Fogarty's Twitter account, shows how she must manage her dialogue with the audience in order to balance her desire to be helpful with her caution about doing too much for a reader.

In a further discussion on this Twitter thread, Fogarty stated that while she enjoys helping students learn and find answers to specific questions, she has learned to be suspicious of inquiries with many small, related questions that are likely to come from a worksheet or some other class assignment (Fogarty 2014).

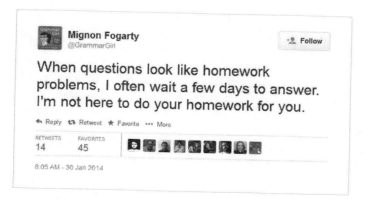

FIGURE 10.2 A Twitter post from Mignon Fogarty in response to a question from a reader (Fogarty 2014).

Fogarty added that she is "extraordinarily grateful" to the community of people who follow the Grammar Girl page and answer other people's questions on a regular basis: "I check the page every day, but I usually find that someone from a core group of four or five people has already adequately answered other people's questions" (pers. comm., November 24, 2013).

Both Doyle and Fogarty believe that their content can support ethical action. Doyle said that it is ethical to employ expert hosts at Quick and Dirty Tips who know their subjects well and will approach them responsibly. The hosts are subject-matter experts in their fields with appropriate academic degrees and other professional designations. Thus, said Doyle, they are bound by their qualifications and experience to provide ethical and accurate information. Doyle gave the example of Nutrition Diva, Monica Reinagel. Doyle said Reinagel gives very practical, science-based advice. She takes time to analyze the research and present a considered case on the issues. Reinagel is a voice of reason in an area where a lot of trends and fads come and go (pers. comm., December 6, 2013). Said Fogarty, "There's so much inaccurate information on the internet. We take great care to ensure that Quick and Dirty Tips articles are written by people who know what they are talking about and that the articles are accurate. I feel that by providing quality information, we can help people make choices that are good, right, and ethical" (pers. comm., November 24, 2013). The Quick and Dirty Tips approach to communication is definitely dialogic. The hosts interact with audience members—online and occasionally in person—to share their expertise and to learn what is important to the audience. By involving audience members and empowering them to give feedback, Quick and Dirty Tips hosts treat listeners and readers as Yous and not Its.

Alan Alda Center for Communicating Science

The Alan Alda Center for Communicating Science at Stony Brook University trains scientists and health professionals to communicate more effectively with the public, elected officials, and others outside their own disciplines. The Alda Center collaborates with Stony Brook University's School of Journalism to offer a variety of workshops and courses on writing, interviewing, and using digital media. One of the Alda Center's hallmarks, however, is using theatrical improvisation exercises and storytelling techniques to help scientists learn how to communicate clearly with nonscientists.

Dr. Christine O'Connell, a trained marine scientist, is a lecturer and workshop coordinator at the Alda Center. She said that actor Alan Alda was instrumental in creating the center that bears his name. Alda has interviewed thousands of scientists over the years for the television show *Scientific American Frontiers* on the Public Broadcasting Service, or PBS. O'Connell said that Alda often had very informative, engaging conversations with the scientists when they would talk off camera. But as soon as the camera turned on, the scientists would switch into lecture mode. O'Connell said it became clear to Alda that scientists need help in having these two-way conversations (pers. comm., December 5, 2013).

O'Connell said that the Alda Center's goal is to help scientists connect to their audiences in clear, vivid, scientifically accurate ways. The goal is not to dumb down the science but to connect to real audiences and to communicate effectively. One main idea that the Alda Center teaches scientists is to know the goal for each communication. Whether the goal is to interest the audience in the subject, get a job, or convince someone to collaborate on a project, the goal should drive the action. O'Connell said that the second main idea is to know the audience: "We talk a lot about how communication is a two-way street. You can't achieve real communication in a lecture; you have to connect to the audience. Part of that two-way communication is to be able to distill the most important parts of the information you have" (pers. comm., December 5, 2013).

O'Connell said that storytelling and improvisation are important skills because both help speakers to connect to the audience. In storytelling and improvisation, speakers have to respond to the feedback that they get from the audience and learn how to change accordingly (pers. comm., December 5, 2013). Valeri Lantz-Gefroh, a longtime acting instructor, is the improvisation coordinator for the Alda Center. She said the public has some misconceptions about improvisation, probably because many people associate improvisation solely with comedy. Lantz-Gefroh said that actors and scholars regard Viola Spolin as the grandmother of improvisation. The Alda Center teaches improvisation in the Viola Spolin tradition as a method of learning to be present and available in the moment of an interpersonal encounter (pers. comm., December 6, 2013).

Lantz-Gefroh said the mirror exercise is one iconic exercise for the Alda Center with many applications. In this exercise, two people face each other. One leads with movement, and the other follows to mirror the leader. Lantz-Gefroh said that just being face-to-face with another person creates some discomfort for the scientists; they often have some giggles to work out. During this exercise, participants look their partners in the eyes for five minutes and soon realize how intimate this is. Lantz-Gefroh said that the partners who are leading this exercise learn to focus on helping the follower keep up rather than on their own actions: "Shifting your focus away from yourself and onto another can transform the choices you make, the words you choose, the body language you use, and the way the message is received. It makes communication a two-way street" (pers. comm., December 6, 2013). In other words, it reinforces how dialogue involves both sending out information and receiving feedback in response.

Scientists, researchers, and other experts are the audience for the Alda Center's workshops and courses. Members of this audience will, in turn, communicate with their own audiences: collaborators, potential collaborators, elected officials, and other members of the general public. Said O'Connell, who participated in the six-week pilot program with Alan Alda, "I am a scientist by training, and I know that it is very hard to learn these techniques. This is not intuitive for scientists.

Scientists are taught to write in third person and to take themselves out of their communication. This process is like trying to relearn communication by putting yourself into it to connect to the audience" (pers. comm., December 5, 2013).

O'Connell said that the techniques Alda used in the pilot program were foreign to her but added that the experience was transformational: "The before and after videos of participants in the program were just amazing. All these scientists transformed how they talked about their research. This communication is really challenging; sometimes I even find it hard to talk with people in my own field because people are so specialized" (pers. comm., December 5, 2013).

O'Connell said that her teaching evaluations improved dramatically after she went through the program. In fact, teaching assistants who go through the program regularly receive significantly better teaching reviews than do those who have not been through it. O'Connell said that two skills that scientists learn through the Alda Center are especially important: "being able to get out of our own heads, and being able to adjust and talk in a different way if we're not reaching the audience. It is a long process to develop these skills, but the results are amazing" (pers. comm., December 5, 2013).

Lantz-Gefroh noted that scientists live passionate lives as they pursue their work; such passion is essential. But they then have to remove that passion when they write their research reports. Scientists who are students at the Alda Center learn to put the human element back into their work when they talk about it. Lantz-Gefroh said they learn to share stories and feelings with the audience because those are things to which nonscientists respond (pers. comm., December 6, 2013).

O'Connell said that dialogue is a core concept at the Alda Center: "We say communication is a two-way street. If you really want to communicate with people, you have to understand them and interact with them. A lecture is one-way communication, and lectures often turn people off." She added that to reach another person, students at the Alda Center have to pay focused attention to that person and not themselves: "This is especially challenging for scientists because we spend so much time in our own heads. You have to get out of your own head and focus on that other person. If you're trying to reach a politician, tell that person why your research matters to them and not just to you" (pers. comm., December 5, 2013).

Both O'Connell and Lantz-Gefroh pointed to ethical impacts of the Alda Center's work. Lantz-Gefroh said faculty at the Alda Center believe that scientists have a responsibility to share the meaning and implications of their work and that an engaged public encourages sound public decision making. She added that the ability to communicate directly and vividly can enhance scientists' career prospects by helping them secure funding, collaborate across disciplines, compete for positions, and serve as effective teachers (pers. comm., December 6, 2013).

O'Connell noted that the general public funds a lot of scientific research through tax dollars that fund grant-giving organizations such as the National

Science Foundation and the National Institutes of Health: "I personally think it is our responsibility to communicate back to the public about what we're doing and why it is important. I think it is part of our civic responsibility to help people understand what we do" (pers. comm., December 5, 2013).

The Alda Center helps scientists learn to be themselves when communicating to nonscientists. Said Lantz-Gefroh,

> We're stuck in a system where politicians and others are "teaching" science while hiding some facts and misrepresenting others. These people have passionate voices, and people respond to that passion. Passionate voices often carry the day. We teach our students to put themselves back into the stories about their work, to avoid jargon, and to select details in a way that communicates the truth accurately and effectively. It's an exciting process. (pers. comm., December 6, 2013)

Many examples of dialogue in this book involve authors learning about their audiences and then adapting to them. While the scientists at the Alda Center certainly do this, they also improve their dialogue skills by learning about themselves.

Conclusion

The approaches described in this chapter show that support for clear communication extends beyond the organizations typically associated with the plain-language movement. The Alda Center shows how theatrical exercises can help scientists, physicians, and other highly skilled experts to "get out of their own heads" and try to understand situations from the audience's point of view. The For Dummies series uses humor and a can-do attitude to discuss technical material that can be dry and sterile. The Quick and Dirty Tips group uses hosts with personas—Grammar Girl, Get-Fit Guy, Nutrition Diva—to create content for audiences they get to know well. Booster Shot Media uses online videos and print materials to help children learn to live with asthma. Common Craft uses online videos with simple visuals and plain-English scripts to provide explanations of many technical subjects. Common Craft and the Quick and Dirty Tips hosts use the internet not only to convey content but also to engage with audiences and share dialogue with them.

Each of these groups challenges the power differential that separates experts from nonexperts (e.g., Walker 2001) empowering consumers to act. Although the groups differ in their approaches, they all show that clear communication in plain language starts with a clear understanding of the audience and respect for the audience's needs for information. Simple, conversational language helps experts meet in the middle on a narrow ridge (Buber 1965) with nonexpert consumers. Each group provides ways for consumers to have dialogue (Buber 1970) about subjects that interest them and affect their lives.

Questions and Exercises

1. Think about a time when you learned an unexpected lesson about communication. Perhaps you completed a theatrical improvisation exercise, like those used at the Alda Center. Perhaps you were a member of a sports team or a musical ensemble. Perhaps you were in a foreign country or were trying to learn a new skill. Describe what you learned from that experience in a memo of 200 to 300 words.

2. Find some online materials from Quick and Dirty Tips and the For Dummies series. Notice the ways they incorporate a friendly and sometimes humorous tone. Using these for inspiration, think of a way to apply a humorous or friendly tone to some content from your organization. In a memo of 300 words, revise two or three paragraphs of your content in a more friendly or humorous style, and then reflect on how such an approach compares with your typical content.

3. Find an online video from Common Craft or Booster Shot Media. Using it for inspiration, create a storyboard for two to three minutes of your own content. Then write a memo of 200 to 300 words analyzing how online videos could add to or repurpose some of the content at your organization.

11
CONCLUSION

In some ways, this book responds to a comment in the preface of a collection published almost three decades ago. As I mentioned in chapter 1, Brockmann (1989b) wrote in *Technical Communication and Ethics*'s preface that "plain language, although a readability concern, is not necessarily an ethical concern. Identification of plain language with ethical language mistakes the outward signs of ethics, plain language, for true ethical actions" (v). In the years since Brockmann's statement, the plain-language movement has grown considerably around the world, and supporters of plain language have cited its many benefits. These developments have presented an opportunity to reexamine Brockmann's assertion. In particular, I have attempted to answer these two questions:

- Is plain language an ethical concern?
- What processes and procedures can help plain-language communicators do ethical work that helps their audiences?

While I agree with Brockmann (as do the professionals in chapter 3) that plain language does not necessarily make a document ethical, I think that evidence presented in the years since Brockmann's pronouncement shows that it is a mistake to omit plain language from discussions of ethics in technical communication. As the statements from plain-language professionals and the plain-language activities profiled in this book have shown, plain language can support ethical action. The ethical nature of plain language lies not in its surface features but in the benefits it can provide in specific contexts—especially BUROC situations.

Plain Language and Ethics

Many sources and people cited in this book, from the professionals in chapter 3 who responded to my questionnaire to those working in the organizations profiled in chapters 4 through 9, indicate that plain language is an ethical concern, a point of view that differs from that of Brockmann (1989b). While plain language is not a panacea for ethical problems, as several plain-language professionals such as Martin Cutts (pers. comm., July 12, 2013) and Rachel McAlpine (pers. comm., June 15, 2013) acknowledged in chapter 3, a connection between plain language and ethics certainly exists. Many plain-language professionals cited in chapter 3, including Audrey Riffenburgh (pers. comm., June 25, 2013) and Sandra Fisher-Martins (pers. comm., July 8, 2013), believe that plain language can be a means of ethical action. The professionals profiled in chapters 4 through 9 believe that as well and reflect that in their work; even some in chapter 10 who do not necessarily identify themselves with the plain-language movement link clear communication with ethics. The usefulness of plain language is clear (e.g., Kimble 2012), and the ethical ideal of utility promotes choosing actions that provide the greatest good for the most people. (Brockmann [1989b] does not discuss the utility of plain language.) Nevertheless, the link between plain language and ethics goes beyond mere utility or effectiveness. The literature on ethics in technical communication and in plain language identifies principles that professionals should follow to treat their audiences with respect and to behave responsibly. Authors in academia and in industry warn against misleading or misguiding audiences (e.g., Walzer 1989a; Kostelnick, 2008), with or without using plain language. The ethics literature also shows a strong regard for individuals' rights (e.g., Kant [1785] 1969; Markel 2001) and for helping people—citizens, consumers, patients, collaborators—exercise their rights. Through Buber's descriptions of I–You dialogic communication, we see that ethical action starts with understanding the audience as You, not It. Dialogic communication helps two parties to communicate more effectively and to understand each other better, and plain language can support this dialogue.

Processes and Procedures for Ethical Plain-Language Work

The projects I have profiled are examples of plain language used for ethical action. Five common threads emerge among the ways they have done their work. These include knowing the audience well, developing content with users, testing content with users, following strong practices for writing and reviewing, and respecting the audience.

Knowing the Audience Well

To create effective plain-language materials, knowing the audience well is critical. Members of the project to restyle the Federal Rules of Evidence know the audience members as their colleagues, students, and clients; through completing law school and working in the legal field for many years, the restyling team for the Evidence Rules has come to know what their audience needs. Civic Design team members writing Field Guides to Ensuring Voter Intent have spent many hours working behind the scenes on elections, ballot design, and voting problems, and they have met many audience members through conferences and usability-testing sessions. This extensive interaction with the audience helps them know the audience's needs. Healthwise employs medical reviewers with significant experience treating patients (some as generalist physicians, some as specialists) and employs many managers and writers who have worked in clinical roles. This extensive experience working with patients informs the decisions Healthwise employees make as they plan, write, and edit content about personal health, wellness, and medical treatments.

Developing Content with Users

User-centered design supports the development of plain-language materials. Kleimann Communication Group combined testing with users and iterative content development to reinvent the mortgage cost disclosure forms mandated by the Truth in Lending Act (TILA) and the Real Estate Settlement Procedures Act (RESPA). Kleimann Communication Group worked with citizens and real estate professionals around the US to collect feedback and incorporate it into the loan estimate and the closing disclosure. The project to restyle the Federal Rules of Evidence included active users of the rules on the project team, and it also included a public comment period in which those outside the project team could provide feedback. The CommonTerms team, which created an HTML generator to provide a preview summary of a company's terms and conditions for the use of its online product, conducted alpha testing with a company offering an online service. The team continues to gather feedback by participating in the Open Notice forum and by inviting potential CommonTerms users to contribute their expertise. These participatory, user-centered approaches show respect to users (Katz and Rhodes 2010).

Testing Content with Users

Testing content with users is a long-standing best practice for plain-language communication; testing complements user-centered design practices that involve users in developing products or content. Kleimann Communication Group conducted extensive user testing while developing integrated mortgage disclosures with the Consumer Financial Protection Bureau. Formative testing collected input from users as the disclosure forms were in development; summative testing revealed that the completed disclosure forms worked well. CommonTerms used a prototype for testing during its alpha phase, and it also conducted a live

trial of a preview document with the Austrian company Gnowsis. This alpha
testing led CommonTerms to revise the name of one of the categories of terms,
and it helped CommonTerms understand how the size and placement of the
"Preview Terms" button affect both users of online systems and the companies
providing those systems. Healthwise conducts formative and summative testing
as it develops and enhances many products across several product lines. In addition to providing a means of quality assurance, testing with users enacts Salvo's
call (2001) for dialogic, ethical interaction between creators and users of systems.

Following Strong Practices for Writing and Reviewing

Plain-language professionals have access to many helpful handbooks and style
guides, and experienced professionals have stored away valuable practical knowledge from their experiences. In addition to these stores of tacit and explicit
knowledge, strong practices for writing and reviewing help ensure the quality of
content and encourage helpful feedback from reviewers. Healthwise uses several
levels of review to ensure that content is both appropriately plain and medically
accurate. The associate editors have authority to edit copy according to Healthwise's plain-language standards, and they participate in content development
early and often. The culture at Healthwise encourages all employees to understand and apply plain-language principles. The processes used to restyle the Federal Rules of Evidence ensured that the style consultant, a plain-language expert,
had authority to lead the work and to influence style choices. The restyling team
followed guidelines specific to its task (Garner 2007) rather than using a general
plain-language guide such as that of Cutts (2009). The division of labor among
the restyling team members defined their clear responsibilities and obligations.
This division of labor helped keep the work on track over a four-year period.
Health Literacy Missouri (HLM) developed its own manual on plain language
and health literacy to use when reviewing client-created documents. HLM then
adapted the content from the manual into a checklist that makes the review process more efficient and more consistent. The manual and the checklist codify the
knowledge that HLM has accumulated, and they help new employees effectively
learn and apply principles of plain language and health literacy.

Respecting the Audience

The practice of respecting the audience does not easily reduce to a set of steps
in a procedure or an activity within a flowchart, but it is present and essential in
each profile of ethical plain-language practice in this book. It entails not merely
involving the audience but demonstrating deep, ongoing concern for the audience's needs for information. Healthwise respects each patient's right and ability
to make decisions about health and medical care; the company creates information in plain language so that patients can use and apply it. Having been involved

with the electoral process for more than a decade, Civic Design respects the process and the people who manage it. The Field Guides to Ensuring Voter Intent give these election officials information they can apply to improve their county's processes and even their ballot designs. The project to restyle the Federal Rules of Evidence respected the audience by encouraging and responding to feedback from audience members over a four-year period. The CommonTerms project respects the concerns of both those who use and those who design software and online services. Through alpha testing, beta testing, and activities of the Open Notice forum that include email discussions and meetings about online privacy and data control, CommonTerms engages many who care about online privacy and data sharing. HLM respects those with low health literacy. It provides several services that help providers and health systems to see health and medical situations from the perspective of people with low health literacy. Instead of ignoring this audience, HLM helps to engage it. Kleimann Communication Group respected the audiences for the TILA–RESPA home-purchase documentation throughout its work on the project. Kleimann and her team started by immersing themselves in the context of home purchases; they reviewed relevant research, discussed technical content with experts, and brainstormed to understand the challenges of creating documents used by real estate and mortgage professionals as well as consumers. They also actively challenged their own expectations of how documents should look; for example, Kleimann assumed that consumers would prefer to see some costs presented in aggregate, but testing showed that consumers preferred a full list of them. Additionally, Kleimann Communication Group sought feedback from a wide range of people involved in home purchases across the US. Audience input affected the TILA–RESPA documents substantially.

The BUROC Model for Identifying Opportunities for Plain Language

Chapter 3 reported responses to the BUROC framework from an international group of plain-language experts. Many of the professionals thought that the BUROC framework—identifying situations that are bureaucratic, unfamiliar, rights oriented, and critical—had good potential to help plain-language writers in their work. Others raised some concerns. Two thought that the BUROC model might discourage a writer from using plain language. Sandra Fisher-Martins, founder of Português Claro, pointed out that plain-language practitioners write to address readers, not situations (pers. comm., July 8, 2013); the BUROC model focuses on situations. Simon Hertnon, managing director of the New Zealand consultancy Nakedize, said, "I would be wary of highlighting any set of situations for fear of providing unintended permission not to employ plain language in other situations" (pers. comm., May 28, 2013). Two others saw problems in the rights-oriented component of the model. Karen Payton said "rights-oriented" can be difficult to understand in its breadth (pers. comm., May 27, 2013).

William DuBay commented, "Just saying or legislating that people have a right will certainly increase awareness of it, but it will not accomplish much and might even give people a false sense of security" about it (pers. comm., May 25, 2013). I share DuBay's reticence to proclaim that individuals have an inherent right to plain language, as in the statement, "Plain language is a civil right." Nevertheless, I believe that plain language is a means to respect individuals' rights; such rights include being respected as an individual and not used as a means for another's ends, as Kant ([1785] 1969) describes in the second formulation of his categorical imperative. Plain language cannot solve all problems people face, and the BUROC framework cannot guarantee that one person will necessarily act ethically toward another. Nevertheless, the BUROC framework can help writers better understand those for whom they write, and it may help them to behave more ethically and more humanely toward people who need information to cope with a situation they face.

The BUROC framework identifies challenging situations that affect people profoundly. It cannot identify all such challenges, but it provides a heuristic to identify many difficult situations and to reflect on how best to help the individuals facing them. Plain-language professionals can use the BUROC framework to better understand the people trying to persevere through those situations. Through a greater understanding of audiences, professionals can better empathize with people facing BUROC situations and provide information to better suit their needs.

Moving Forward: Plain Language and Ethics

At the 2013 conference of Plain Language Association International (PLAIN), I had the privilege of participating in a panel discussion on ethics and plain language. One of my fellow panelists, Renée Sarojini Saklikar, encouraged plain-language professionals to consider facets of their audience members' lived experiences that the BUROC model does not explicitly address, including gender, race, ethnicity, culture, history, and economic status. Saklikar encouraged plain-language professionals to involve audiences when creating messages for them. She invoked a saying shared often by people who feel marginalized and unrepresented in matters that affect them, such as people with disabilities or individuals from First Nations or other native groups: nothing about us without us (Saklikar 2013). That is, groups pushed to society's margins or governed without adequate representation deserve to participate in dialogue about the treatment they receive. Like all citizens, marginalized and underrepresented people deserve authentic dialogue that involves them as Yous, not technical dialogue that orders them around as Its.

Dialogue is not always possible or practical, but it is an ethical ideal for which to strive continually. In BUROC situations, individuals often struggle against a bureaucracy, whether an insurance company, a government agency, or even a medical provider. They often feel treated as Its and not Yous; they experience the bureaucracy talking at them or down to them; they wish someone would speak to them, in dialogue, with mutual respect. They wish they could meet the bureaucracy in the middle, on a "narrow ridge" (Buber 1965) where a representative of the bureaucracy will listen. Listening is essential to dialogue—dialogue demands both speaking and listening. The foundation of dialogue is mutual respect between parties.

Plain-language professionals should take care to respect the people for whom they write and to involve them in content development whenever possible. The organizations profiled in this book have done ethical plain-language work by both respecting and working with their audiences in a number of ways. By treating their audiences as Yous and not Its, enacting Buber's (1970) dialogic ethics, plain-language professionals act ethically. Simply producing information that is ostensibly clear and well organized does constitute a dialogue with the audience. Plain-language communicators enact dialogic ethics by engaging their audiences and getting to know them well, by understanding the circumstances that separate the audience from the individual or organization attempting to communicate, by working to establish common ground between communicator and audience, and by understanding that communicators and audiences function best in a mutual, reciprocal relationship.

REFERENCES

Agresti, Alan, and Brett Presnell. 2002. "Misvotes, Undervotes, and Overvotes: The 2000 Presidential Election in Florida." *Statistical Science* 17 (4): 436–40.

Ali, Nadia. 2012. "Community Participation: A Powerful Resource for Removing Health Literacy Barriers in Health Care Organizations." *Engaging the Patient*, October 25. Accessed February 12, 2013. http://engagingthepatient.com/2012/10/25/community-participation-a-powerful-resource-for-removing-health-literacy-barriers-in-health-care-organizations/.

Allen, Lori, and Dan Voss. 1997. *Ethics in Technical Communication: Shades of Gray*. New York: Wiley.

———. 1998. "Ethics for Editors: An Analytical Decision-Making Process." *IEEE Transactions on Professional Communication* 41 (1): 58–65.

Allen, Nancy. 1996. "Ethics and Visual Rhetorics: Seeing's Not Believing Anymore." *Technical Communication Quarterly* 5 (1): 87–105.

Amare, Nicole, and Alan Manning. 2013. *A Unified Theory of Information Design: Visuals, Text, & Ethics*. Amityville, NY: Baywood.

Aristotle. 1975. *The Nichomachean Ethics*. Translated by H. Rackham. Cambridge, MA: Harvard University Press.

Arkestål, Hanna, and Carl Törnquist. 2012. "Alla tackar ja." *CommonTerms*. Accessed November 24, 2013. http://commonterms.net/alpha/userstudy2012.pdf.

Arnett, Ronald C. 1986. *Conversation and Community: Implications of Martin Buber's Dialogue*. Carbondale and Edwardsville: Southern Illinois University Press.

———. 2004. "A Dialogic Ethic 'between' Buber and Levinas: A Responsive Ethical 'I.'" In *Dialogue: Theorizing Difference in Communication Studies*, edited by Rob Anderson, Leslie A. Baxter, and Kenneth N. Cissna, 75–90. Thousand Oaks, CA: Sage.

Ballentine, Brian. 2008. "Professional Communication and a 'Whole New Mind': Engaging with Ethics, Intellectual Property, Design, and Globalization." *IEEE Transactions on Professional Communication* 51 (3): 328–40.

Baron, Marcia. 2011. "Virtue Ethics in Relation to Kantian Ethics: An Opinionated Overview and Commentary." In *Perfecting Virtue: New Essays on Kantian Ethics and Virtue*

Ethics, edited by Lawrence Jost and Julian Wuerth, 8–37. Cambridge, UK: Cambridge University Press.

Bowden, Peta. 1997. *Caring: Gender-Sensitive Ethics*. London and New York: Routledge.

Bowen, Betsy A., Thomas M. Duffy, and Erwin R. Steinberg. 1991. "Analyzing the Various Approaches of Plain Language Laws." In *Plain Language: Principles and Practice*, edited by Erwin R. Steinberg, 19–29. Detroit: Wayne State University Press.

Boyatzis, Richard E. 1998. *Transforming Qualitative Information*. Thousand Oaks, CA: Sage.

Brockmann, R. John. 1989a. "A Historical Consideration of Ethics and the Technical Writer: From the 1880's to the 1980's." In *Technical Communication and Ethics*, edited by R. John Brockmann and Fern Rook, 107–12. Washington, DC: Society for Technical Communication.

———. 1989b. "What Is This Collection About?" In *Technical Communication and Ethics*, edited by R. John Brockmann and Fern Rook, v–vi. Washington, DC: Society for Technical Communication.

———. 1994. "Gregory Clark's Take on Technical Communication Ethics: Flimsy, Fragile, but Correct." *Journal of Computer Documentation* 18 (3): 11–15.

Brockmann, R. John, and Fern Rook, eds. 1989. *Technical Communication and Ethics*. Washington, DC: Society for Technical Communication.

Bryan, John. 1992. "Down the Slippery Slope: Ethics and the Technical Writer as Marketer." *Technical Communication Quarterly* 1 (1): 73–88.

———. 1995. "Seven Types of Distortion: A Taxonomy of Manipulative Techniques Used in Charts and Graphs." *Journal of Technical Writing and Communication* 25 (2): 127–79.

Buber, Martin. 1965. *Between Man and Man*. New York: Macmillan.

———. 1967. "Hope for This Hour." In *The Human Dialogue: Perspectives on Communication*, edited by Floyd W. Matson and Ashley Montagu, 306–12. New York: Free Press.

———. 1970. *I and Thou*. Translated by Walter Kaufmann. New York: Charles Scribner's Sons.

Buchholz, William James. 1989. "Deciphering Professional Codes of Ethics." *IEEE Transactions on Professional Communication* 32 (2): 62–68.

Center for Plain Language. 2014. "Center for Plain Language | 2014 ClearMark Winners." *Center for Plain Language*. Accessed May 2, 2014. http://centerforplainlanguage.org/awards/2014-clearmark-winners/.

Cheek, Annetta. 2012. "The Plain Regulations Act, HR 3786." *Michigan Bar Journal* 91 (5): 40–41.

Chisnell, Dana E. 2013. "Civic Design." Accessed October 21, 2013. http://civicdesigning.org/.

Clark, Gregory. 1987. "Ethics in Technical Communication: A Rhetorical Perspective." *IEEE Transactions in Professional Communication* 30 (3): 190–95.

———. 1990. *Dialogue, Dialectic, and Conversation: A Social Perspective on the Function of Writing*. Carbondale and Edwardsville: Southern Illinois University Press.

———. 1994. "Professional Ethics from an Academic Perspective." *Journal of Computer Documentation* 18 (3): 32–38.

Colton, David, and Robert W. Covert. 2007. *Designing and Constructing Instruments for Social Research and Evaluation*. San Francisco: Jossey-Bass.

Common Craft. 2014. "Our Story." *Common Craft*. Accessed September 9, 2014. http://www.commoncraft.com/our-story.

CommonTerms. 2013a "CommonTerms—About the Project." *CommonTerms*. Accessed November 12. http://commonterms.net/About.aspx.

1

ographyReferences **185**

———. 2013b. "CommonTerms—About the Project—Roadmap." *CommonTerms.* Accessed November 26. http://commonterms.net/Roadmap.aspx.

———. 2013c. "CommonTerms—What Is the Problem?" *CommonTerms.* Accessed November 9. http://www.commonterms.net/Problem.aspx.

———. 2013d. "CommonTerms—What Others Did." *CommonTerms.* Accessed November 25. http://commonterms.net/Related.aspx.

Consumer Financial Protection Bureau. 2013a. "Integrated Mortgage Disclosures under the Real Estate Settlement Procedures Act (Regulation X) and the Truth in Lending Act (Regulation Z)." *Know Before You Owe.* Last modified November 20. Accessed December 1, 2013. http://files.consumerfinance.gov/f/201311_cfpb_final-rule_integrated-mortgage-disclosures.pdf.

———. 2013b. "Timeline." *Know Before You Owe.* Accessed December 1, 2013. http://www.consumerfinance.gov/knowbeforeyouowe/timeline/.

Ctrl-Shift. 2013. "Transparency about the Use of Personal Data." *Ctrl-Shift News.* Last modified October 11. Accessed November 26, 2013. https://www.ctrl-shift.co.uk/news/2013/10/11/transparency-about-the-use-of-personal-data/.

Cutts, Martin. 2009. *Oxford Guide to Plain English.* 3rd ed. New York: Oxford University Press.

Davis, Michael. 1994. "The Rhetoric of Ethics Compromised." *Journal of Computer Documentation* 18 (3): 15–19.

Dieterich, Dan, Mary Bowman, and Sarah Pogell. 2006. "The Writing Coach's Perspective on Workplace Writing: A Conversation with Lee Clark Johns." *Issues in Writing* 16 (2): 103–22.

Doheny-Farina, Stephen. 1987. "Special Issue on Legal and Ethical Aspects of Technical Communication." *IEEE Transactions on Professional Communication* 30 (3): 119–20.

———. 1989. "Ethics and Technical Communication." In *Technical and Business Communication: Bibliographic Essays for Teachers and Corporate Trainers,* edited by Charles H. Sides, 53–69. Urbana, IL: National Council of Teachers of English.

Dombrowski, Paul M. 1995. "Can Ethics Be Technologized? Lessons from Challenger, Philosophy, and Rhetoric." *IEEE Transactions on Professional Communication* 38 (3): 146–50.

———. 2000a. *Ethics in Technical Communication.* Boston: Allyn and Bacon.

———. 2000b. "Ethics and Technical Communication: The Past Quarter Century." *Journal of Technical Writing and Communication* 30 (1): 3–29.

———. 2007. "The Evolving Face of Ethics in Technical and Professional Communication: Challenger to Columbia." *IEEE Transactions on Professional Communication* 50 (4): 306–19.

Dorney, Jacqueline M. 1988. "The Plain English Movement." *English Journal* 77 (3): 49–51.

Douglas, Davison M., Sidney A. Fitzwater, Daniel J. Capra, Robert A. Hinkle, Joseph Kimble, Joan N. Ericksen, Marilyn L. Huff, Reena A. Raggi, Geraldine Soat Brown, Edward H. Cooper, Kenneth S. Broun, Harris L. Hartz, Katharine Traylor Schaffzin, Roger C. Park, Deborah J. Merritt, Andrew D. Hurwitz, W. Jeremy Counseller, and Paula Hannaford-Agor. 2012. "Symposium: The Restyled Federal Rules of Evidence." *William and Mary Law Review* 53:1435–547.

Dragga, Sam. 1996. "'Is This Ethical?' A Survey of Opinion on Principles and Practices of Document Design." *Technical Communication* 43 (3): 255–65.

———. 1997. "A Question of Ethics: Lessons from Technical Communicators on the Job." *Technical Communication Quarterly* 6 (2): 161–78.

———. 2011. "Cooperation or Compliance: Building Dialogic Codes of Conduct." *Technical Communication* 58 (1): 4–18.

Dragga, Sam, and Dan Voss. 2001. "Cruel Pies: The Inhumanity of Technical Illustrations." *Technical Communication* 48 (3): 265–74.

———. 2003. "Hiding Humanity: Visual and Verbal Ethics in Accident Reports." *Technical Communication* 50 (1): 61–82.

DuBay, William H. 2004. "The Principles of Readability." *Plain Language at Work Newsletter*, August 25. Accessed February 8, 2013. http://www.impact-information.com/impactinfo/readability02.pdf.

Dumas, Joseph S., and Janice C. Redish. 1999. *A Practical Guide to Usability Testing*. Revised ed. Portland, OR: Intellect Books.

Ede, Lisa, and Andrea Lunsford. 1990. *Singular Texts/Plural Authors: Perspectives on Collaborative Writing*. Carbondale and Edwardsville: Southern Illinois University Press.

Ehn, Pelle. 1993. "Scandinavian Design: On Participation and Skill." In *Participatory Design: Principles and Practices*, edited by Douglas Schuler and Namioka Aki, 41–77. Hillsdale, NJ: Lawrence Erlbaum Associates.

Ericsson, Mikael, and Pär Lannerö. 1997. *Internetboken*. Stockholm: Statens Skolverk.

Faber, Brenton. 1999. "Intuitive Ethics: Understanding and Critiquing the Role of Intuition in Ethical Decisions." *Technical Communication Quarterly* 8 (2): 189–202.

Felker, Daniel B., ed. 1980. *Document Design: A Review of the Relevant Research*. Washington, DC: American Institutes for Research.

Felker, Daniel B., Frances Pickering, Veda R. Charrow, V. Melissa Holland, and Janice C. Redish. 1981. *Guidelines for Document Designers*. Washington, DC: American Institutes for Research.

Felsenfeld, Carl. 1991. "The Plain English Experience in New York." In *Plain Language: Principles and Practice*, edited by Erwin R. Steinberg, 13–18. Detroit: Wayne State University Press.

Fogarty, Mignon. 2014. "Twitter/GrammarGirl: When questions look like homework . . ." Twitter post. January 30, 8:05 a.m. Accessed January 30, 2014. https://twitter.com/GrammarGirl/status/42892187243290624.

Foley, Edward B. 2011. "The Lake Wobegone Recount: Minnesota's Disputed 2008 U.S. Senate Election." *Election Law Journal* 10 (2): 129–64.

Friedman, Maurice S. 1955. *Martin Buber: The Life of Dialogue*. Chicago: University of Chicago Press.

Garner, Bryan A. 2007. "Guidelines for Drafting and Editing Court Rules." *United States Courts*. Accessed November 4, 2013. http://www.uscourts.gov/uscourts/RulesAndPolicies/rules/guide.pdf.

Gilligan, Carol. 1982. *In a Different Voice: Psychological Theory and Women's Development*. Cambridge, MA: Harvard University Press.

Girill, T.R. 1994. "Achieving Principled Communication amid Conflict." *Journal of Computer Documentation* 18 (3): 19–24.

Gnecchi, M., B. Maylath, B. Mousten, F. Scarpa, and S. Vandepitte. 2011. "Field Convergence between Technical Writers and Technical Translators: Consequences for Training Institutions." *IEEE Transactions on Professional Communication* 54 (2): 168–84.

Graves, Heather, and Roger Graves. 2011. *A Strategic Guide to Technical Communication*. 2nd ed. Peterborough, ON: Broadview.

Greer, Rachelle R. 2012. "Introducing Plain Language Principles to Business Communication Students." *Business Communication Quarterly* 75 (2): 136–52.

Griffin, Jack. 1989. "When Do Rhetorical Choices Become Ethical Choices?" In *Technical Communication and Ethics*, edited by R. John Brockmann and Fern Rook, 63–70. Arlington, VA: Society for Technical Communication.

Hall, Dean G., and Bonnie A. Nelson. 1987. "Integrating Professional Ethics into the Technical Writing Course." *Journal of Technical Writing and Communication* 17 (1): 45–62.

Harner, Sandra W., and Tom G. Zimmerman. 2002. *Technical Marketing Communication.* New York: Pearson Longman.

Hawthorne, Mark D. 2001. "Learning by Doing: Teaching Decision Making through Building a Code of Ethics." *Technical Communication Quarterly* 10 (3): 341–55.

Health Literacy Missouri. 2014a. "Health Environment Assessments." *Health Literacy Missouri.* Accessed April 1. http://www.healthliteracymissouri.org/Our-Services/Health-Environment-Assessments.

———. 2014b. "Health Literacy Missouri Plain Language." *Vimeo.* Accessed May 1. http://vimeo.com/91523960.

———. 2014c. "Learn the Facts." *Health Literacy Missouri.* Accessed April 1. http://www.healthliteracymissouri.org/About-Us/Learn-the-Facts.

Healthwise, Inc. 2014. "A Force for Good: Healthwise and the Informed Medical Decisions Foundation Merge." *Healthwise.* Accessed May 2, 2014. http://www.healthwise.org/insights/about/specialpages/forceforgood.aspx.

Herrington, TyAnna K. 1995. "Ethics and Graphic Design: A Rhetorical Analysis of the Document Design in the 'Report of the Department of the Treasury on the Bureau of Alcohol, Tobacco, and Firearms Investigation of Vernon Wayne Howell also Known as David Koresh.'" *IEEE Transactions on Professional Communication* 38 (3): 151–57.

———. 2003. *A Legal Primer for a Digital Age.* New York: Pearson Longman.

Hertnon, Simon. 2011. "Simon Hertnon's Theory of Universal Human Needs." *Nakedize.* Last modified May 11. Accessed July 1, 2013. http://www.nakedize.com/universal-human-needs.cfm.

House, Richard, Anneliese Watt, and Julia M. Williams. 2004. "Teaching Enron: The Rhetoric and Ethics of Whistle-Blowing." *IEEE Transactions on Professional Communication* 47 (4): 244–55.

Instagram. 2013. "Terms of Use." Last modified January 19. Accessed September 7, 2014. http://instagram.com/legal/terms/#.

International Consortium for Clear Communication. 2011. "Project." Accessed May 1, 2013. http://icclear.net/project/.

John Wiley & Sons. 2013. "About For Dummies." *Dummies.com.* Accessed December 6, 2013. http://www.dummies.com/about-for-dummies.html.

Johnson, Robert R. 1998. *User-Centered Technology: A Rhetorical Theory for Computers and Other Mundane Artifacts.* Albany: State University of New York Press.

Jones, Dan. 1998. *Technical Writing Style.* Boston: Allyn and Bacon.

Kallendorf, Craig, and Carol Kallendorf. 1989. "Aristotle and the Ethics of Business Communication." *Journal of Business and Technical Communication* 3 (1): 54–69.

Kant, Immanuel. (1785) 1969. "Foundations of the Metaphysics of Morals." In *Kant: Foundations of the Metaphysics of Morals: Text and Critical Essays*, edited by Robert P. Wolff, translated by L. W. Beck. Indianapolis, IN: Bobbs-Merrill.

Katz, Steven B. 1992. "The Ethic of Expediency: Classical Rhetoric, Technology, and the Holocaust." *College English* 54 (3): 255–75.

———. 1993. "Aristotle's Rhetoric, Hitler's Program, and the Ideological Problem of Praxis, Power, and Professional Discourse." *Journal of Business and Technical Communication* 7 (1): 37–62.

Katz, Steven B., and Vicki W. Rhodes. 2010. "Beyond Ethical Frames of Technical Relations: Digital Being in the Workplace World." In *Digital Literacy for Technical Communication: 21st Century Theory and Practice*, edited by Rachel Spilka, 230–56. New York and London: Routledge.

Kienzler, Donna, and Carol David. 2003. "After Enron: Integrating Ethics into the Professional Communication Curriculum." *Journal of Business and Technical Communication* 17 (4): 474–89.

Kimble, Joseph. 2006. *Lifting the Fog of Legalese: Essays on Plain Language.* Durham, NC: Carolina Academic Press.

———. 2009. "Another Example from the Proposed New Federal Rules of Evidence." *Michigan Bar Journal* 88 (9): 46–48.

———. 2012. *Writing for Dollars, Writing to Please.* Durham, NC: Carolina Academic Press.

———. 2013. "Wild and Crazy Tales from a Decade of Drafting US Federal Court Rules." Paper presented at the biennial meeting of Plain Language Association International, Vancouver, British Columbia, October 10.

Kleimann Communication Group. 2012. "Know Before You Owe: Evolution of the Integrated TILA-RESPA Disclosures." *Know Before You Owe.* Last modified July 9. Accessed November 12, 2013. http://www.consumerfinance.gov/f/201207_cfpb_report_tila-respa-testing.pdf.

———. 2013a. "Know Before You Owe: Post-Proposal Consumer Testing of the Spanish and Refinance Integrated TILA-RESPA Disclosures." *Know Before You Owe.* Last modified November 20. Accessed December 4, 2013. http://files.consumerfinance.gov/f/201311_cfpb_report_tila-respa_testing-spanish-refinancing.pdf.

———. 2013b. "Know Before You Owe: Quantitative Study of the Current and Integrated TILA-RESPA Disclosures." *Know Before You Owe.* Last modified November 20. Accessed December 1, 2013. http://files.consumerfinance.gov/f/201311_cfpb_study_tila-respa_disclosure-comparison.pdf.

Kostelnick, Charles. 2008. "The Visual Rhetoric of Data Displays: The Conundrum of Clarity." *IEEE Transactions on Professional Communication* 51 (1): 116–30.

LaDuc, Linda M. 1997. "From Schroedinger's Cat to Flaming on the Internet: Exploring Gender's Relevance for Technical/Professional Communication." In *Foundations for Teaching Technical Communication: Theory, Practice, and Program Design*, edited by Katherine Staples and Cezar Ornatowski, 119–32. Greenwich, CT: Ablex Publishing.

LaDuc, Linda, and Amanda Goldrick-Jones. 1994. "The Critical Eye, the Gendered Lens, and 'Situated' Insights—Feminist Contributions to Professional Communication." *Technical Communication Quarterly* 3 (3): 245–56.

Lancaster, Amber. 2006. "Rethinking Our Use of Humanistic Aspects: Effects of Technical Communication beyond the Intended Audience." *Technical Communication* 53 (2): 212–24.

Lannerö, Pär. 2011. "What Do You Accept? Common Terms of Service Proposal." *Pär Lannerös blog.* Last modified April 8. Accessed November 9, 2013. http://p.lannero.com/cats/.

———. 2012. "Previewing Online Terms and Conditions: CommonTerms Alpha Proposal." *CommonTerms.* Last modified January 27. Accessed November 12, 2013. http://commonterms.net/commonterms_alpha_proposal.pdf.

———. 2013. "Fighting the Biggest Lie on the Internet: CommonTerms Beta Proposal." Last modified April 30. Accessed October 26, 2013. http://www.commonterms.net/commonterms_beta_proposal.pdf.

Lauchman, Richard. 2010. *Punctuation at Work: Simple Principles for Achieving Clarity and Style*. New York: AMACOM.

Lay, Mary M. 1994. "Feminist Theory and the Redefinition of Technical Communication." In *Humanistic Aspects of Technical Communication*, edited by Paul M. Dombrowski, 141–59. Amityville, NY: Baywood.

LeFever, Lee. 2010. "RSS in Plain English: Three Years Old Today." *The CommonCraft Blog*. Last modified April 22. Accessed December 4, 2013. http://www.commoncraft.com/rss-plain-english-three-years-old-today.

———. 2012. *The Art of Explanation*. New York: Wiley.

Lefler, Dion. 2013. "Senate Approves Bill for Plain-Language 'Explainer' on Complicated Ballot Issues." *The Wichita Eagle*. March 25. Accessed May 1, 2013. http://www.kansas.com/2013/03/25/2732942/senate-approves-bill-for-plain.html#storylink=misearch.

Locke, Joanne. 2004. "A History of Plain Language in the United States Government." *Plainlanguage.gov*. Accessed January 25, 2013. http://www.plainlanguage.gov/whatisPL/history/locke.cfm.

Longaker, Mark Garrett. 2005. "Beyond Ethics: Notes toward a Historical Materialist Paideia in the Professional Writing Classroom." *Journal of Business and Technical Communication* 19 (1): 78–97.

Lowy, Joan. 2008. "Jet Fuel-Tank Protection Ordered." *SeattlePI.com*, July 18. Accessed December 10, 2013. http://www.seattlepi.com/business/article/Jet-fuel-tank-protection-ordered-1279529.php.

Lutz, William. 1988. "Fourteen Years of Doublespeak." *English Journal* 77 (3): 40–42.

Mack, John, and Anna Wittel. 2001. *The New Frontier: Exploring eHealth Ethics*. Newtown, PA: Internet Healthcare Coalition.

Mackinnon, Jamie. 1997. "Ethics and Plain Language: A Rhetorical Perspective." Paper presented at the biennial meeting of Plain Language Consultants Network, Calgary, Alberta, September 24–26. Accessed November 8, 2013. http://www.plainlanguagenetwork.org/conferences/1997/#NINE.

Malone, Edward A. 2011. "The First Wave (1953–1961) of the Professionalization Movement in Technical Communication." *Technical Communication* 58 (4): 285–306.

Markel, Mike. 1991. "A Basic Unit on Ethics for Technical Communicators." *Journal of Technical Writing and Communication* 21 (4): 327–50.

———. 1993. "An Ethical Imperative for Technical Communicators." *IEEE Transactions on Professional Communication* 36 (2): 81–86.

———. 1997. "Ethics and Technical Communication: A Case for Foundational Approaches." *IEEE Transactions on Professional Communication* 40 (4): 284–98.

———. 2001. *Ethics in Technical Communication: A Critique and Synthesis*. Westport, CT: Ablex Publishing.

Maylath, Bruce. 1997a. "Why Do They Get It When I Say 'Gingivitis' but Not When I Say 'Gum Swelling'?" In *Approaches to Teaching Non-Native English Speakers across the Curriculum*, edited by David L. Sigsbee, Bruce W. Speck, and Bruce Maylath, 29–37. San Francisco: Jossey-Bass.

———. 1997b. "Writing Globally: Teaching the Technical Writing Student to Prepare Documents for Translation." *Journal of Business and Technical Communication* 11 (3): 339–52.

Maylath, Bruce, and Emily A. Thrush. 2000. "Café, thé, ou Lait? Teaching technical communicators to manage translation and localization." In *Managing Global Communication in Science and Technology*, edited by Peter J. Hager and H. J. Scheiber, 233–54. New York: John Wiley & Sons.

Mazur, Beth. 2000. "Revisiting Plain Language." *Technical Communication* 47 (2): 205–11.

McArthur, Tom. 1991. "The Pedigree of Plain English." *English Today* 7 (3): 13–19.

McDonald, Aleecia M., and Lorrie Faith Cranor. 2008. "The Cost of Reading Privacy Policies." *I/S: A Journal of Law and Policy for the Information Society* 4 (3): 540–65.

McLaughlin, G. Harry. 1969. "SMOG Grading—A New Readability Formula." *Journal of Reading* 12 (8): 639–46.

Mendelson, Michael. 1993. "A Dialogical Model for Business Correspondence." *Journal of Business and Technical Communication* 7 (3): 283–311.

Michaelson, Herbert. 1990. "How an Author Can Avoid the Pitfalls of Practical Ethics." *IEEE Transactions on Professional Communication* 33 (2): 58–61.

Mill, John Stuart. 1863. *Utilitarianism.* London: Parker, Son, and Bourn.

Miller, Carolyn R. 1979. "A Humanistic Rationale for Technical Writing." *College English* 40 (6): 610–17.

Mowat, Christine. 1999. *A Plain Language Handbook for Legal Writers.* Toronto: Carswell.

Murray, Elspeth. 2006. "This Is Bad Enough." *ElspethMurray.com.* January 20. Accessed April 4, 2013. http://www.elspethmurray.com/Poems/poems_badenough.htm.

National Council of Teachers of English. 2009. "Shining a Spotlight on Doublespeak." Accessed May 1, 2013. http://archives.library.illinois.edu/ncte/about/september.php.

Nelson-Burns, Carol. 2004. "The Davis-Besse Nuclear Power Plant Eroded Reactor Head: A Case Study." *IEEE Transactions on Professional Communication* 47 (4): 268–80.

Nielsen, Jakob. 2000. "Why You Only Need to Test with 5 Users." *Nielsen Norman Group.* Accessed September 10, 2014. http://www.nngroup.com/articles/why-you-only-need-to-test-with-5-users/.

———. 2012. "How Many Test Users in a Usability Study?" *Nielsen Norman Group.* Accessed September 10, 2014. http://www.nngroup.com/articles/how-many-test-users/.

Noble, H. B. 1999. "Hailed as a Surgeon General, Koop Is Faulted on Web Ethics." *New York Times,* September 5. Accessed April 13, 2013. http://search.proquest.com/docview/43 1246969?accountid=9649.

Noddings, Nel. 2003. *Caring: A Feminine Approach to Ethics and Moral Education.* 2nd ed. Berkeley: University of California Press.

Open Notice. 2013 "Open Notice » About." *Open Notice.* Accessed November 25. http://opennotice.org/about/.

Ornatowski, Cezar M. 1992. "Between Efficiency and Politics: Rhetoric and Ethics in Technical Writing." *Technical Communication Quarterly* 1 (1): 91–103.

———. 1997. "Technical Communication and Rhetoric." In *Foundations for Teaching Technical Communication: Theory, Practice, and Program Design*, edited by Katherine Staples and Cezar Ornatowski, 31–51. Greenwich, CT: Ablex Publishing.

Orwell, George. 2005. "Politics and the English language." In *Why I Write*, by George Orwell, 102–20. New York: Penguin.

Osborne, Helen. 2005. *Health Literacy from A to Z: Practical Ways to Communicate Your Health Message.* Sudbury, MA: Jones and Bartlett Publishers.

Perica, Louis. 1972. "Honesty in Technical Communication." *Technical Communication* 15: 2–6.

Plain Language Action and Information Network (PLAIN). 2013a. "About Us." Accessed May 1. http://www.plainlanguage.gov/site/about.cfm.

———. 2013b. "Plain Writing Act of 2010: Federal Agency Requirements." Accessed May 1. http://www.plainlanguage.gov/plLaw/law/index.cfm.

Porter, James E. 1987. "Truth in Technical Advertising: A Case Study." *IEEE Transactions in Professional Communication* 30 (3): 182–89.

———. 1993a. "Developing a Postmodern Ethics of Rhetoric and Composition." In *Defining the New Rhetorics*, edited by Theresa Enos and Stuart C. Brown, 207–26. Newbury Park, CA: Sage.

———. 1993b. "The Role of Law, Policy, and Ethics in Corporate Composing: Toward a Practical Ethics for Professional Writing." In *Professional Communication: The Social Perspective*, edited by Nancy Roundy Blyler and Charlotte Thralls, 128–43. Newbury Park, CA: Sage.

Possin, Kevin. 1991. "Ethical Argumentation." *Journal of Technical Writing and Communication* 21 (1): 65–72.

Radez, Frank. 1989. "STC and the Professional Ethic." In *Technical Communication and Ethics*, edited by R. John Brockmann and Fern Rook, 3–5. Arlington, VA: Society for Technical Communication.

Redish, Janice C. 1985. "The Plain English Movement." In *The English Language Today*, edited by Sidney Greenbaum, 125–38. Oxford: Pergamon Press.

Rentz, Kathryn C., and Mary Beth Debs. 1987. "Language and Corporate Values: Teaching Ethics in Business Writing Courses." *Journal of Business Communication* 24 (3): 37–48.

Republic of South Africa. 2014. "Constitution of the Republic of South Africa, 1996." *South African Government*. Accessed October 24, 2014. http://www.gov.za/documents/constitution-republic-south-africa-1996.

Riley, Kathryn. 1993. "Telling More than the Truth: Implicature, Speech Acts, and Ethics in Professional Communication." *Journal of Business Ethics* 12 (3): 179–96.

Riley, Kathryn, and Jo Mackiewicz. 2003. "Globalizing Plain English: Can Plain Be Polite?" *The World within the Words: Business Practice in Plain Language*, edited by Clive Muir, 13–22. Toronto, Canada: Association for Business Communication.

Riley, Kathryn, Michael Davis, Apryl Cox Jackson, and James Maciukenas. 2009. "'Ethics in the Details': Communicating Engineering Ethics via Micro-Insertion." *IEEE Transactions on Professional Communication* 52 (1): 95–108.

Russell, David R. 1993. "The Ethics of Teaching Ethics in Professional Communication: The Case of Engineering Publicity at MIT in the 1920s." *Journal of Business and Technical Communication* 7 (1): 84–111.

Sachs, Harley. 1989. "Ethics and the Technical Communicator." In *Technical Communication and Ethics*, edited by R. John Brockmann and Fern Rook, 7–10. Arlington, VA: Society for Technical Communication.

Saklikar, Renée Sarojini. 2013. "Ethics in Communication: Understanding Situations for Plain Language Use." Panelist at the biennial meeting of Plain Language Association International, Vancouver, British Columbia, October 12.

Salvo, Michael J. 2001. "Ethics of Engagement: User-Centered Design and Rhetorical Methodology." *Technical Communication Quarterly* 10 (3): 273–90.

Sanders, Scott P. 1988. "How Can Technical Writing Be Persuasive?" In *Solving Problems in Technical Writing*, edited by Lynn Beene and Peter White, 55–78. New York: Oxford University Press.

———. 1997. "Technical Communication and Ethics." In *Foundations for Teaching Technical Communication: Theory, Practice, and Program Design*, edited by Katherine Staples and Cezar Ornatowski, 99–117. Greenwich, CT: Ablex Publishing.

Sarkos, Constantine. 2011. "Improvements in Aircraft Fire Safety Derived from FAA Research over the Last Decade." *Federal Aviation Administration* May. http://www.fire.tc.faa.gov/pdf/TN11–8.pdf.

Sauer, Beverly A. 1993. "Sense and Sensibility in Technical Documentation: How Feminist Interpretation Strategies Can Save Lives in the Nation's Mines." *Journal of Business and Technical Communication* 7 (1): 63–83.

Sawyer, Thomas M. 1988. "The Argument about Ethics, Fairness, or Right and Wrong." *Journal of Technical Writing and Communication* 18 (4): 367–75.

Schriver, Karen A. 1997. *Dynamics in Document Design: Creating Texts for Readers*. New York: John Wiley & Sons.

Schroll, Christopher J. 1995. "Technology and Communication Ethics: An Evaluative Framework." *Technical Communication Quarterly* 4 (2): 147–64.

Scott, J. Blake. 1995. "Sophistic Ethics in the Technical Writing Classroom: Teaching *Nomos*, Deliberation, and Action." *Technical Communication Quarterly* 4 (2): 187–99.

Shimberg, H. Lee. 1989a. "Ethics and Rhetoric in Technical Writing." In *Technical Communication and Ethics*, edited by R. John Brockmann and Fern Rook, 59–62. Arlington, VA: Society for Technical Communication.

———. 1989b. "Technical Communicators and Moral Ethics." In *Technical Communication and Ethics*, edited by R. John Brockmann and Fern Rook, 11–13. Arlington, VA: Society for Technical Communication.

Sims, Brenda R. 1993. "Linking Ethics and Language in the Technical Communication Classroom." *Technical Communication Quarterly* 2 (3): 285–99.

Sladek, Sarah L. 2011. *The End of Membership as We Know It: Building the Fortune-Flipping, Must-Have Association of the Next Century*. Washington, DC: ASAE.

Society for Technical Communication (STC). 2012a. "About STC." Accessed October 3, 2012. http://www.stc.org/about-stc/the-society/.

———. 2012b. "Ethical Principles." Accessed October 3, 2012. http://www.stc.org/about-stc/the-profession-all-about-technical-communication/ethical-principles.

———. 2012c. "Mission & Vision." Accessed October 3, 2012. http://www.stc.org/about-stc/the-society/mission-vision.

Steinberg, Erwin R. 1991a. "Introduction: Promoting Plain Language." In *Plain Language: Principles and Practice*, edited by Erwin R. Steinberg, 7–10. Detroit: Wayne State University Press.

———, ed. 1991b. *Plain Language: Principles and Practice*. Detroit: Wayne State University Press.

Strother, Judith B. 2004. "Crisis Communication Put to the Test: The Case of Two Airlines on 9/11." *IEEE Transactions on Professional Communication* 47 (4): 290–300.

Sturges, David L. 1992. "Overcoming the Ethical Dilemma: Communication Decisions in the Ethic Ecosystem." *IEEE Transactions on Professional Communication* 35 (1): 44–50.

Sullivan, Dale L. 1990. "Political-Ethical Implications of Defining Technical Communication as a Practice." *Journal of Advanced Composition* 10 (2): 375–86.

Sullivan, Dale L., and Michael S. Martin. 2001. "Habit Formation and Story Telling: A Theory for Guiding Ethical Action." *Technical Communication Quarterly* 10 (3): 251–72.

Sullivan, Patricia, and James E. Porter. 1997. *Opening Spaces: Writing Technologies and Critical Research Practices*. Greenwich, CT: Ablex Publishing.

Swanton, Christine. 2003. *Virtue Ethics: A Pluralistic View*. Oxford: Oxford University Press.

Tebeaux, Elizabeth. 1997. *The Emergence of a Tradition: Technical Writing in the English Renaissance, 1475–1640*. Amityville, NY: Baywood.

Thrush, E. A. 2001. "Plain English? A Study of Plain English Vocabulary and International Audiences." *Technical Communication* 48 (3): 289–96.

Tillery, D. 2005. "The Plain Style in the Seventeenth Century: Gender and the History of Scientific Discourse." *Journal of Technical Writing and Communication* 35 (3): 273–89.

Voss, Dan, and Madelyn Flammia. 2007. "Ethical and Intercultural Challenges for Technical Communicators and Managers in a Shrinking Global Marketplace." *Technical Communication* 54 (1): 72–87.

Wagstaff, Keith. 2012. "You'd Need 76 Work Days to Read All Your Privacy Policies Each Year." *TIME.com*. March 6. Accessed November 2, 2013. http://techland.time.com/2012/03/06/youd-need-76-work-days-to-read-all-your-privacy-policies-each-year/.

Walker, Margaret Urban. 2001. "Seeing Power in Morality: A Proposal for Feminist Naturalism in Ethics." In *Feminists Doing Ethics*, edited by Peggy DesAutels and Joanne Waugh, 3–14. Lanham, MD: Rowman & Littlefield.

Walzer, Arthur E. 1989a. "The Ethics of False Implicature in Technical and Professional Writing Courses." *Journal of Technical Writing and Communication* 19 (2): 149–60.

———. 1989b. "Professional Ethics, Codes of Conduct, and the Society for Technical Communication." In *Technical Communication and Ethics*, edited by R. John Brockmann and Fern Rook, 101–5. Arlington, VA: Society for Technical Communication.

———. 1994. "Ethical Norms for Technical Communication: Plato, Aristotle, and Clark's 'Rhetorical Perspective.'" *Journal of Computer Documentation* 18 (3): 25–32.

Wand, Jonathan N., Kenneth W. Shotts, Jasjeet S. Sekhon, Jr., Walter R. Mebane, Michael C. Herron, and Henry E. Brady. 2001. "The Butterfly Did It: The Aberrant Vote for Buchanan in Palm Beach County, Florida." *The American Political Science Review* 95 (4): 793–810.

Ward Sr., Mark. 2010. "The Ethic of Exigence: Information Design, Postmodern Ethics, and the Holocaust." *Journal of Business and Technical Communication* 24 (1): 60–90.

Whitehouse, Roger. 1999. "The Uniqueness of Individual Perception." In *Information Design*, edited by Robert Jacobson, 103–29. Cambridge, MA: MIT Press.

Wicclair, Mark R., and David K. Farkas. 1989. "Ethical Reasoning in Technical Communication: A Practical Framework." In *Technical Communication and Ethics*, edited by R. John Brockmann and Fern Rook, 21–25. Arlington, VA: Society for Technical Communication.

Williams, David E. 2007. "'Grammar Girl' a Quick and Dirty Success." *CNN.com*. Accessed December 6, 2013. http://www.cnn.com/2007/TECH/internet/01/22/grammar.girl/index.html.

Zoetewey, Meredith W., and Julie Staggers. 2004. "Teaching the Air Midwest Case: A Stakeholder Approach." *IEEE Transactions in Professional Communication* 47 (4): 233–43.

INDEX